Medical Intelligence Unit

TNF Superfamily

Sanjay Khare
Amgen Inc.
Thousand Oaks, California, U.S.A.

CRC Press
Taylor & Francis Group
Boca Raton London New York

CRC Press is an imprint of the
Taylor & Francis Group, an **informa** business

TNF Superfamily

Medical Intelligence Unit

CRC Press
Taylor & Francis Group
6000 Broken Sound Parkway NW, Suite 300
Boca Raton, FL 33487-2742

First issued in hardback 2019

ISBN 13: 978-1-58706-306-0 (hbk)

Visit the Taylor & Francis Web site at
http://www.taylorandfrancis.com

and the CRC Press Web site at
http://www.crcpress.com

Library of Congress Cataloging-in-Publication Data

TNF superfamily / [edited by] Sanjay Khare.
p. ; cm. -- (Medical intellegence unit)
Includes bibliographical references.
ISBN-13: 978-1-58706-306-0
1. Tumor necrosis factor. I. Khare, Sanjay. II. Series: Medical intelligence unit (Unnumbered : 2003)
[DNLM: 1. Tumor Necrosis Factors. QU 55.2 T6265 2007]
QR185.8.T84T56 2007
616.07'9--dc22

2007034203

CONTENTS

EDITOR

Sanjay Khare
Amgen Inc.
Thousand Oaks, California, U.S.A.

CONTRIBUTORS

Pratima Bansal-Pakala
Division of Immunochemistry
La Jolla Institute for Allergy and Immunology
San Diego, California, U.S.A.
Chapter 2

Dale E. Bredesen
Buck Institute for Age Research
Novato, California, U.S.A.
Chapter 4

Lieping Chen
Department of Dermatology
Johns Hopkins University School of Medicine
Baltimore, Maryland, U.S.A.
Chapter 6

Michael Croft
Division of Immunochemistry
La Jolla Institute for Allergy and Immunology
San Diego, California, U.S.A
Chapter 2

Yang-Xin Fu
Department of Pathology
The University of Chicago
Chicago, Illinois, U.S.A.
Chapter 5

Iqbal S. Grewal
Preclinical Therapeutics
Seattle Genetics, Inc.
Bothell, Washington, U.S.A.
Chapter 1

Ji Li
Department of Metabolic Disorders
Amgen Inc.
One Amgen Center Drive
Thousand Oaks, California
Chapter 3

Rammohan V. Rao
Buck Institute for Age Research
Novato, California
and
University of California, San Francisco
San Francisco, California, U.S.A.
Chapter 4

Shahram Salek-Ardakani
Division of Immunochemistry
La Jolla Institute for Allergy and Immunology
San Diego, California, U.S.A.
Chapter 2

Takanori So
Division of Immunochemistry
La Jolla Institute for Allergy and Immunology
San Diego, California, U.S.A.
Chapter 2

Jianxun Song
Division of Immunochemistry
La Jolla Institute for Allergy and Immunology
San Diego, California, U.S.A.
Chapter 2

Jing Wang
Department of Pathology
and Committee on Immunology
The University of Chicago
Chicago, Illinois, U.S.A.
Chapter 5

Ryan A. Wilcox
Department of Immunology
Mayo Clinic
Rochester, Minnesota, U.S.A.
Chapter 6

PREFACE

The tumor necrosis factor/receptor [TNF/TNFR] superfamily consists of more than 20 transmembrane proteins with conserved N-terminal cysteine-rich domains [CRDs] in the extracellular ligand binding region. Members have wide tissue distribution and play important roles in biological processes such as lymphoid and neuronal development, innate and adaptive immune response, and cellular homeostasis.

The chapters of this book address some of the most interesting functions of the TNF/TNFR superfamily.

In Chapter 1 Iqbal Grewal describes the CD40 cell surface receptor and its ligand CD154 [CD40L] which mediate contact-dependent signals between B and T cells. The chapter emphasizes major and newly discovered findings for the roles of CD40-CD154 in the cellular differentiation, survival and death pathways in lymphoid organogenesis and in the activation and homeostasis of immune cells.

OX40 [CD134] and its ligand OX40L are members of the TNF/TNFR superfamily. As described in Chapter 2 by Michael Croft and colleagues, they have been shown to be crucial for T cell-mediated immune reactions, specifically for T cell costimulation.

Skeletal homeostasis is maintained by a balance between bone-resorbing osteoclasts and bone-building osteoblasts. Three TNF ligand and receptor family members have been identified as crucial in the extracellular regulation of bone resorption: osteoprotegerin [OPG] receptor activator of NK-κB ligand [RANKL] or osteoprotegerin ligand [OPGL] and receptor activator of NF-κB. The new model of the regulation of osteoclastogenesis and bone resorption involving these TNF superfamily members is discussed in Chapter 3 by Ji Li.

Several proinflammatory cytokines, notably TNF-α, have been shown to mediate diverse forms of experimental neurodegeneration, and both neurotoxic and neuroprotective actions have been reported. In Chapter 4 Rao and Bredesen review evidence for the role of cytokines in neurodegeneration in general and for the role of TNF in particular.

LIGHT, a newly discovered TNF superfamily member [TNFSF14], is a type II transmembrane protein expressed on activated T cells and immature dendritic cells. In Chapter 5 Wang and Fu describe the role of LIGHT in the induction of autoimmunity. LIGHT and LTab share the same receptor, LTbR, and cooperate in lymphoid organogenesis and development of lymphoid structure.

CD137 is a member of the TNF receptor superfamily that can be induced on a variety of cells, including activated T lymphocytes, natural killer cells and dendritic cells. In Chapter 5 Wilcox and Chen describe studies that suggest that the CD137 activation pathway is capable of regulating cellular and molecular components of both innate and adaptive immunity.

Overall, the TNF/TNFR superfamily is remarkable for its ability to induce effects either on cell survival or apoptotic cell death.

Chapter 1

CD40 and CD154

Iqbal S. Grewal*

Abstract

CD40 is a cell surface receptor that belongs to the tumor necrosis factor (TNF) receptor superfamily and was initially shown to be critical for mediating contact-dependent signals between B and T cells. Thus, early studies have established that interactions of CD40 with its ligand, CD154 (CD40L), a TNF superfamily member are essential for generation of thymus dependent (TD) humoral immune responses. Recent studies demonstrate that CD40 is not only expressed on B cells but also on dendritic cells, follicular dendritic cells, monocytes, macrophages, mast cells, fibroblasts and endothelial cells. Likewise, expression of CD154 has also recently been demonstrated on many cells types beyond T cells as originally thought. Studies with CD40- and CD154-knockout mice and blocking antibodies to interfere with CD40-CD154 interaction have elucidated the role of CD40-CD154 in many aspects of the immune system. These studies have established a key role of CD40-CD154 in regulation of immunity and autoimmunity and have firmly established the basis for many preclinical and clinical investigations. This article focuses on major and newly discovered findings for a role of CD40-CD154 in the regulation of many functions in the immune system and applications of this system for a therapeutic utility.

Introduction

In the past several years many molecules that provide stable intercellular contacts, costimulatory or apoptotic signals essential for functioning of the immune system have been identified, and molecular mechanisms of their functioning have been elucidated. About two decades ago CD40 was discovered as a cell surface molecule expressed on B cells during all stages of development, which interacts with its ligand, CD154 (CD40, gp39, T-BAM, TRAP) primarily expressed on activated $CD4^+$ T cells. Use of antibodies that disrupt interactions between CD40-CD154 has established a clear role of this molecular interaction in TD humoral response.[1,2] Role of these interactions in humans was established by the

*Iqbal S. Grewal—Seattle Genetics, Inc., Preclinical Therapeutics, 21823 30th Drive SE, Bothell, WA 98021, USA. Email: igrewal@seagen.com

TNF Superfamily, edited by Sanjay Khare. ©2007 Landes Bioscience

discovery of mutation in CD154, which lead to the development of a severe form of X-linked immunodeficiency, hyper IgM syndrome (HIGM1), characterized by the inability to mount TD humoral response.[3,4] As a result, HIGM1 patients suffer from recurrent pathogenic bacterial infections.[5] A similar form of deficiency was also found in CD40- and CD154-deficient mice.[6-8]

It is clear now that CD40 expression is not only restricted to B cells; other types of cells such as dendritic cells, follicular dendritic cells (DC), monocytes, macrophages, mast cells, fibroblasts and endothelial cells also express CD40.[9,10] It is also true now that expression of CD154 is not only restricted to activated $CD4^+$ T cells; other cell types and platelets also express CD154.[11] These findings suggest that CD40-CD154 interactions play a much broader role in diverse aspects of the immune response (reviewed in refs. 9,10). Thus, a key role of this interaction in innate immunity, priming and expansion of antigen-specific CD4 T cells, activation of DC to become competent antigen presenting cells (APC) and cross priming of cytotoxic T cells (CTL), activation of macrophages, and activation of endothelia have now been established.[9-16] CD40-CD154 interactions are involved in generation, amplification and effector aspects of immune cells important in many inflammatory situations. Thus, blocking CD40-CD154 interactions by using anti-CD154 antibodies or CD40- and CD154- knockout mice have proven importance of CD40-CD154 in transplantation, autoimmunity, atherosclerosis, Alzheimer's disease and infection models in mice.

Since CD40 and CD154 molecules are expressed on cell surfaces, they are thought to be mediating their effects via cognate interaction. CD154 is also found in secreted form as a soluble molecule, which can induce its effects at distant sites from its production. Both CD40 and CD154 belong to TNF superfamily. Members of this family are characterized by strong structural homology, common signaling pathways, and, diverse biological functions such as cell differentiation, activation and cell death. CD40-CD154 pair clearly manifests the properties of typical TNF family members. A number of reviews have discussed many aspects of CD40-CD154 interaction in detail, especially their role in humoral responses.[9,10,17,18] Considering the regulation of broad cellular functions by CD40-CD154 in addition to their well established role in humoral responses, the present article will highlight developments in our understanding of the importance of CD40-CD154 interactions in nonB cell mediated cellular functions.

CD40/CD154 Structural Features and Their Expression

A human CD40 gene was originally cloned from Burkitt lymphoma Raji cell line.[19] The mature molecule is a 48 kDa transmembrane glycoprotein cell surface receptor. CD40 protein consists of 277 amino acid residues with 193 amino acid residues in extracellular domain, 21 amino acid residues leader sequence and 22 amino acid residues transmembrane domain. CD40 molecule consists of 62 amino acid residues long cytoplasmic tail, which serves to activate several pathways through adaptor molecules binding.[20] CD40 is a typical type I transmembrane protein, and human CD40 gene is expressed as a single 1.5 kb mRNA transcript. The murine form of the CD40 protein consists of 305 amino acid residues with 193 amino acid residues in the extracellular domain, 21 amino acid residues in the leader sequence, 22 amino acid residues in the transmembrane domain and a 90 amino acid residues long cytoplasmic tail. Human and murine form of the molecule share 62% amino acid residues and last 32 C-terminal amino acid residues are conserved. In the extracellular domain, 22/22 cysteine amino acid residues are conserved suggesting murine and human molecules structurally similar in folding. Due to alternative splicing murine CD40 is expressed in two mRNA transcripts of sizes 1.7 and 1.4 kb. CD40 is expressed on all developmental stages of B cells, dendritic cells, follicular dendritic cells, monocytes, macrophages, mast cells, CD8 T cells, hematopoietic progenitors, fibroblasts, endothelial cells, malignant B cells and many carcinomas.[9,10,21]

Murine CD154 gene was originally cloned from thymoma cell line, EL-4, which was mapped to X-chromosome.[22] The mature molecule is a 33 kDa transmembrane glycoprotein cell surface ligand. CD154 protein consists of 260 amino acid residues with 214 amino acid residues in extracellular domain, 24 amino acid residues in the transmembrane domain and 22 amino acid residues long cytoplasmic tail.[10] CD154 is a typical type II transmembrane protein, lacking any detectable signaling motifs on short cytoplasmic tail and expressed as an extracellular carboxy terminus protein. The gene for human CD154 is also mapped to X-chromosome and mutation of this gene lead to X-linked immunodeficiency, HIGM1. Human CD154 gene encodes a protein consists of 261 amino acid residues, 22 amino acid residues long cytoplasmic tail, a 24 amino acid residues transmembrane domain and 215 amino acid residues extracellular domain. Human and murine forms of the molecules share 78% amino acid residues. In the extracellular domain there is a single N-linked glycosylation site in both mouse and human CD154. CD154 exist both as membrane bound trimer and a shorter soluble trimer form. The structure of CD154 has been resolved by X-ray crystallography, and three-dimensional organization is similar to TNFα and LTα proteins.[23] CD154 is predominantly expressed on activated CD4 T cells and platelets, but is also expressed at low levels on activated B cells, NK cells, monocytes, dendritic cells, endothelial cells and smooth muscle cells.[10,24,25] Expression of CD154 on T cells is very tightly regulated and is transient. Upon activation of T cells, expression of CD154 can be seen within 1-2 hours and maximal expression at 6-8 hours with a gradual loss of molecule due to proteolytic cleavage of the surface CD154 and down regulation of CD154 mRNA.

Signalling through CD40

A number of binding motifs have been identified in the cytoplasmic tail of the CD40 molecule; these include binding motifs for kinases such as p23 and Jak3 and TRAFs such as TRAF2, TRAF3 and TRAF6. CD40 signal transduction is very complex which involves different mediators and pathways. Thus, many studies have demonstrated involvement of serine/ threonine kinases such as JNK/SAPK, P38,

MAPK and ERK in CD40 signalling. Crosslinking of CD40 on cells results in activation of NF-κB and other transcription factors. CD40 induces NF-κB activation mediated by degradation of IκBα and IκBβ. Thus, activation of CD40 exerts important biological effects on various cells expressing CD40. For example, CD40 activation on B-cells leads to isotype class switching of immunoglobulin (Ig) genes and affinity maturation of antibodies.[2]

Regulation of Activity of APC

Interactions between T cells and APC are required for the initiation of a successful T cell mediated immune response. APC provide necessary signals for proper activation of the T cells and for maintaining specificity of the T cell response. The main signal delivered to T cell is via interaction of its T cell receptor (TCR) with MHC/peptide complex on the surface of the APC, which maintains specificity and other signals come from the costimulatory molecules on the surface of APC triggering coreceptors on T cells for full activation. Resting B cells which express low levels costimulatory molecules are poor APCs and require activation first in order to upregulate costimulatory activity and to become competent APC.[26] CD154 expressing T cells have been demonstrated to activate resting B cells to progress them through the cell cycle,[27] to upregulate IL-2, IL-4 and IL-5 receptors,[28,29] costimulatory ligands and to deliver costimulatory activity. CD40-CD154 interactions are also important in the induction of costimulatory activity on other types of APC such as DC and macrophages.[9,10] Ligation of CD40 with CD154 on the surface of DC is shown to regulate the production pro-inflammatory cytokines, such as IL-8, MIP-1α, TNFα and IL-12.[30-33] Production of these mediators by DCs is important for inflammatory response, and production of IL-12 by DCs in particular is essential for development of Th1 type responses. Stimulation of human peripheral monocytes via CD40 enhances survival of monocytes and leads to upregulation of CD54, MHC class II, CD86 and CD40 molecules important for APC functions. Thus, regulation of activation of APC via CD40-CD154 interactions is critical step in T cell activation.

Role of CD40-CD154 Interaction in T Cell Priming

The fact that human HIGM1 patients succumb to opportunistic pathogens, such as *Pneumocystis* and *Cryptosporidium,* which typically associate with severe T cell immunodeficiency such as AIDS indicate that T cell responses are defective in the absence of CD154.[5] This defect is also reproduced in CD154 knockout mice, where severe defects in priming of CD4 T cells to protein antigens are seen.[9,13] Recently, a requirement for CD40 signal in T cell priming to alloantigens was also demonstrated, suggesting the importance of CD40-CD154 interactions in priming of alloreactive T cells.[34] Similarly, demonstration of defective immune response to autologous tumors in CD154-deficient mice also reinforces the critical role of CD40-CD154 in T cell priming.[34] A lack of CD4

T cell priming to adenovirus infection was also observed in CD154-deficient mice.[15] As costimulatory signals are important for activation of T cells, lack of priming and expansion of CD4 T cells to protein antigens and to alloantigens in the absence of CD154 was explained by the inability of the CD154-deficient T cells to appropriately activate APC. APCs such as DCs are generally required for initiation of in vivo antigen specific T cell responses, lack of activation of DCs in CD40 dependent manner may explain defects seen in CD4 T cell priming in the above systems.[30,33]

Importance of CD40-CD154 in CD4 T cell priming is well established, however, the priming of CD8 T cells seems to be independent of this interaction. Activation of primary CD8 cytotoxic T cells (CTLs) following infection of CD154-deficient mice with lymphocytic choriomeningitis virus (LCMV), Pichinde virus (PV) or vesicular stomatitis virus (VSV), indicate that priming of CD8 T cells is independent of CD40-CD154.[35-37] However, anti-viral CD8 CTL memory responses were defective in CD154-deficient mice,[35] suggesting a requirement of CD154 mediated signal in the establishment or maintenance of CTL memory. Since CD154 is important in CD4 T cell priming, defect in the memory responses in CD154-deficient mice could be due to inefficient CD4 T cell help. The precise role of CD154, however, in CTL memory response still remains to be established.

CD40-CD154 and Inflammation

Monocytes and macrophages are key players in the T cell mediated inflammatory response; they mediate tissue damage, serve as APCs for T cells, and are effector cells to eliminate intracellular pathogens. Ligation of CD40 on monocytes is important in production of IL-1α, IL-1β, TNFα, IL-6 and IL-8, as well as in the rescue of circulating monocytes from apoptotic death,[38,39] suggesting that CD40-CD154 interactions play an important role at the sites of inflammation. A role of CD40-CD154 interaction was also postulated in the activation of macrophages to produce IFNγ, NO and IL-12.[39-42] Pro-inflammatory cytokines such as, IL-12 is important for Th1 type immune responses and NK cell activation, thus, CD40 dependent macrophage activation may be a key step in inflammatory response.

Since expression of CD40 is seen on endothelial cells from spleen, skin, thyroid gland, thymus and lungs and is upregulated by cytokines,[43-47] a role for CD40-CD154 interactions has been suggested in activation of vascular endothelium. Ligation of CD40 on endothelial cells has been shown to induce CD62E, CD106 and CD54. These molecules are involved in extravasation of leukocytes to the sites of inflammation and secondary lymphoid organs.[44-49] CD40-CD154 interactions also induce adhesion molecules and chemokines in endothelial cells, thus, it is conceivable that CD40 promotes extravasation and accumulation of activated T cells at the site of inflammation. These activated T cells can further activate vascular endothelium, which could then perpetuate inflammation.

CD40-CD154 and Atherosclerosis

Atherosclerosis is an inflammatory disease of vascular system. Involvement of activated T cells, endothelial cells, smooth muscle cells and macrophages in atherosclerotic plaques is well documented.[50-54] All key cell types involved in plaque formation express CD40 and are susceptible for potent CD154 signals. Indeed, recent studies have confirmed an import part played by CD40-CD154 in atherosclerosis. CD154-deficient mice also mutant for ApoE have impairment in the development of early to advanced plaque formation.[55,56] Treatment of atherosclerosis prone mice with blocking antibodies to CD154 also show stable plaque phenotype and much reduced number of infiltrating T cells and macrophages in the plaque area, suggesting a role for CD40-CD154 in atherosclerosis.[57] Treatment of hyperlipidemic low-density lipoprotein receptor-deficient mice, manifesting initial plaques, when treated with a CD154 blocking antibody for three months showed considerable reduction in atherosclerotic lesion area again reinforcing strong role of CD40-CD154 in atherosclerosis.[58] The efficacy seen with the use of anti-CD154 antibody in animal models of atherosclerosis clearly suggests CD40/CD154 as a potential therapeutic target for human disease.

Role of CD40-CD154 in Autoimmunity

A potential role of CD40-CD154 in development of autoimmunity has been addressed using blocking anti-CD154 antibodies in many recent reports. Thus, studies have indicated the importance of CD40-CD154 interactions in many organ-specific T cell mediated autoimmune diseases in mouse models, such as collagen induced arthritis, experimental allergic encephalomyelitis (EAE), lupus nephritis, Type I diabetes, colitis and oophoritis.[59,60,63] Treatment of mice with anti-CD154 antibodies was shown to have profound effect on disease development in all the above mentioned animal models, which was accompanied by a reduction or elimination of damage to the target tissue or infiltration of leukocytes to the target tissues. In addition, CD154-deficient mice expressing MBP-TCR also have been investigated for the development of antigen induced EAE.[14] In these mice, development of EAE was completely blocked. One possible explanation of prevention of development of autoimmunity is that priming of self-antigen specific T cells is inhibited by anti-CD154 or by lack of CD154 expression. On the other hand, these antibodies may also block downstream roles of CD154 on effector functions that mediate tissue damage. As discussed earlier, CD154 regulates both the initiation of CD4 T cell responses and effector functions, such as the activation of macrophages. The efficacy seen in models for lupus, arthritis and oophoritis could also be explained by strong inhibition of autoantibody production by anti-CD154 or lack of CD154 expression. Thus, it is likely that CD40-CD154 regulate both the initiation and effector phases of the responses as well as transmigration of activated T cells to target tissues during an autoimmune response.

Interestingly, in human patients of SLE an increased expression of CD154 is seen in circulating T cells and many CD154 positive cells are found in kidney biopsies,[64,65] suggesting a potential role for CD40-CD154 in human autoimmune disease. Increased expression of CD40 and CD154 is also seen in CNS sections from multiple sclerosis patients.[66] These studies suggest that blockade of CD40-CD154 in ongoing disease may be possible in humans as has been found efficacious in late stage disease models of SLE and EAE in animals.[62,66-68]

Role of CD40-CD154 Interactions in Transplantation

Since CD40-CD154 interactions are critical for priming of T cells to alloantigens,[18,34] their role in transplantation has also been extensively investigated. A potential role of CD40-CD154 in generation of T regulatory cells and in induction of tolerance has also been suggested.[69] Thus, blockade of CD40-CD154 by using anti-CD54 antibodies has been investigated to prolong the survival of solid organ transplants including both allografts and xenografts. These studies included transplantation of skin, heart and allo-islet transplants, and results from these studies have shown that administration of anti-CD154 antibodies could prolong graft survival to some degree, but blockade of the cosimulatory molecules were also required for better survival.[70-72] Prolonged survival of cardiac allografts from anti-CD154 antibody treated mice also showed decreased expression of transcripts for the macrophage products suggesting that in cardiac allo-graft rejection, CD40-CD154 interactions mediate effector functions such as macrophage activation. However, when anti-CD154 antibodies were administered in combination with CTLA-4-Ig, cardiac allografts survival was accompanied by an inhibition of T cell associated cytokine transcripts for IL-2, IL-4, IL-10 and IFNγ suggesting regulation of both T cell and macrophage functions by CD40-CD154.[71] Surprisingly, beneficial effects of anti-CD154 in kidney transplantation in rhesus monkey model either alone or in combination with CTLA-4 have been seen, suggesting the importance of regimen treatment to achieve efficacy.[73,74]

CD40-CD154 in Innate Immune Response

A potential role for CD40 receptor in the innate immune system was recently proposed, where it was described that complement fragment 4b (C4b) binding protein, a regulator of the classical complement pathway can bind CD40 on the surface of B cells and efficiently activate them.[75,76] Thus, this mechanism suggests a novel way by which CD40 can potentially act to link innate and adaptive immune responses. Since C4b binding protein is regulated by inflammatory cytokines, it was proposed that during innate immune response CD40 pathway could be influenced. A lack of C4b binding protein to B cells from CD40-deficient humans and failure to activate B cells from patients who have mutations in CD40 signaling pathway, IKKγ/NEMO, was demonstrated.[75] Just like CD154, in the presence of IL-4 C4b binding protein was able to induce immunoglobulin class switching to IgE, suggesting C4b binding protein directly activate B cells in the absence of CD154. C4b binding protein does not block binding of CD154 to B cells, suggesting that CD154 and C4b binding protein bind to distinct sites on CD40, and thus, potentially could synergize with each other to affect B cell functions. Since both CD40 and C4b binding protein are known to bind bacterial products, it is possible that they may link innate and adaptive immune response to bacteria and other pathogens. Although these initial findings are very exciting, there still remain many important unanswered questions, for example can C4b binding protein activate any other cell type expressing CD40 such as DCs, macrophages, endothelial cells and CD8 T cells?

A potential role for platelet derived CD154 is also proposed in linking innate immune system and adaptive immune components.[76] It is well established that platelets function to modulate local inflammatory events through the release of chemokines, cytokines, and a number of immune-modulatory ligands, including CD154. A potential role of platelet-derived CD154 in modulation of adaptive immunity was recently described.[76] It was demonstrated that platelets, via CD154, induce dendritic cell maturation, B cell isotype switching, and augment CD8 T cell responses both in vitro and in vivo. Since platelets can influence early innate immune response, the data presented above suggest a potential role of platelets via CD154 in initiation of early anti-microbial response. Thus, it was proposed that following early inflammatory and traumatic events associated with invasion, by secreting CD154 platelets form a link between innate and adaptive immune response.[76] In this scenario platelets can transmit early signals to control homeostasis of B cells and DCs via CD154, which could ultimately translate into T cell-mediated adaptive response. Although a role of platelet derived CD154 is proposed in the innate immune system, at this time it is not clear how pathogen-associated molecular patterns, CpG dinucleotides and cellular stress induced factors such as TNF and IL-1 synergize with CD154 to mount an innate immune response. Undoubtedly, this will be the topic of extensive research in forthcoming years.

Role of CD40 Ligand in Amyloidosis and Alzheimer's Disease

Recent studies have indicated that Alzheimer's disease has a substantial inflammatory component, and activated microglia play a central role in neuronal degeneration.[78] Since freshly solublized amyloid-beta peptides induce CD40 expression on cultured microglia, and microglia from a transgenic murine model of Alzheimer's disease express higher level of CD40, studies were initiated to investigate the potential role of CD40-CD154 in Alzheimer's disease. Results from these studies show that CD154 induces microglial activation in response to amyloid-beta peptide, which is associated with Alzheimer's disease-like neuronal tau hyperphosphorylation in vivo.[78] Transgenic mice overproducing this peptide, but deficient in CD154, show decreased astrocytosis and microgliosis associated with diminished amyloid-beta peptide levels and beta-amyloid plaque load.[79] Furthermore, by using antibody against CD154 in transgenic mouse model of Alzheimer's disease a marked attenuation of disease pathology, which was accompanied with decreased amyloidogenic processing of amyloid precursor protein and increased circulating levels of amyloid-beta peptide.[80] These findings suggest an important role for CD40-CD154 interactions in mechanisms underlying Alzheimer's disease pathology. These studies also validate the CD40-CD154 interaction as a target for therapeutic intervention in Alzheimer's disease.

CD40-CD154 in the Control of Infection

A potential role of CD40-CD154 interactions in a broad spectrum of anti-infective host immune responses in infections with bacteria, parasites or viruses has been extensively studied (reviewed in ref. 9). CD40-CD154 interactions were found to be critical for the development of humoral responses to extracellular bacterial pathogens, which are resolved mainly by humoral immune mechanisms. Considering a critical role of CD40-CD154 in T cell activation and T cell help to B cells, a role of CD40-CD154 is expected for anti-bacterial immune responses where TD antibodies are important, as also evident by increase susceptibility of HIGIM1 patients to certain microbial infections. HIGM1 patients are susceptible to *Pneumocystis carinii* infection, an opportunistic pathogen of the lungs, and both humoral and cell mediated immune responses are required for the resolution of this infection.[81] Use of anti-CD154 antibodies in *Pneumocystis* infections in mouse model have been shown to completely block the recovery of mice from infection, suggesting a role of CD40-CD154 in anti-infective immunity.[82] This protection was dependent on activation of macrophages for resolution of *Pneumocystis* infections.[81]

Importance of CD40-CD154 has also been demonstrated in the resolution of infection that requires both humoral and cellular immune responses such as infections with multicellular extracellular pathogens *Heligmosomoides polygyrus*. Treatment of mice with anti-CD154 antibody inhibits both cellular and humoral immune responses to *H. polygyrus* infection by downregulating IgG1 and IgE levels, blood eosinophils, intestinal mucosal mast cells and cytokines, such as IL-3, IL-4, IL-5 and IL-9.[83]

Studies in CD40- and CD154-deficient mice[84,85] demonstrate a critical role of CD40-CD154 interactions in the protective immune response to *Leishmania major* because control mice remained resistant to *L. major* promastigote challenge, whereas both CD40- and CD154-deficient mice were unable to control growth of the parasite and developed ulcerating lesions at the site of inoculation. The inability of these mice to clear pathogens was determined to be due to lack of IFNγ production and priming of Th1 type cells. A partial protection from *L. major* infection could be conferred by administration of IL-12,[84] implying that T cells in CD40- and CD154-deficient mice were defective in inducing IL-12 production in macrophages. Infection of CD154-deficient mice with another species, *Leishmania amazonensis* also results in the development of progressive ulcerative lesions that fail to resolve.[86] As expected, defective activation of macrophages was responsible for the lack of protection. Many studies now document that CD40-meditated signals are required for the production of cytokines and NO by macrophages.[38,39] Taken together these results indicate that CD40-CD154 interactions are required for priming of T cells, the differentiation of Th1 effector cells via IL-12 and thus the production of cytokines is required for macrophage activation.

Role of CD40-CD154 in Host Defense against Virus Infections

As CD40-CD154 interactions are critical for humoral and CD4 T cell immune responses, it is anticipated that CD40-CD154 interactions are required for some viral infections where these responses are required, but not for those where CD8 CTL responses are sufficient to control infection. CD154-deficient mice that were infected with LCMV, PV and LSV viruses mount an efficient primary antiviral CTL response, which rapidly clears virus from the host.[35-37] CD154-deficient mice, however, were defective in memory CTL response against these viruses,[35] suggesting the importance of CD40-CD154 interactions in CTL memory responses. As expected, in these mice, TD immune responses were compromised, resulting in the lack of germinal center formation, a short-lived serum titers of virus-specific antibodies, and poor B cell memory.[35-37] Similarly, a role for CD40-CD154 interactions was also studied by infecting CD154-deficient mice with an adenoviral virus vector.[15,37] As this infection is controlled by both CTL and humoral immune responses that are dependent on CD4 T cell help,[87] infection of CD154-deficient mice with this viral vector resulted in poor anti-viral CTL and TD responses suggesting CD40-CD154 interactions are of critical importance in this system as well.

Role of CD40-CD154 in HIV Infections

Many studies have documented that circulating mononuclear cells from HIV-infected individuals produce lower levels of IL-2 and IL-12, and have defective lymphocyte proliferation even when CD4 counts are normal.[88-90] This defective immunity ultimately leads to opportunistic infections in HIV-infected individuals. Blood cells from seropositive HIV-infected individuals when stimulated in vitro with infective pathogen *Toxoplasma gondii* show defective IL-12 production.[88-90] As human IL-12 production in response to *T. gondii* is CD154 dependent,[91-93] expression of CD154 on T cells in HIV-infected patients was investigated. Results from these studies indicated impaired expression of CD154 in HIV-infected patients regardless of their CD4 counts or anti-viral therapy.[94] Furthermore, in vitro activation of T cells also results in a lower expression of CD154 in HIV-infected individuals.[94] The fact that anti-viral therapies do not completely reconstitute immune response,[95,96] studies presented above indicate that altered levels of CD154 levels may contribute immunodeficiency in HIV-infected individuals.

CD40-CD154 Regulate Immune Response at Multiple Levels

It is evident now that for an effective immune response, interactions involving a variety of cell types and multiple surface molecules are required, and data reviewed here suggest that CD40-CD154 interactions are critical for the development of many aspects of this response. Experiments discussed here demonstrate that CD40-CD154 interactions mediate both CD4 T cell dependent and independent cell mediated immune response at multiple levels. A model that could be proposed for a role for CD40-CD154 in the immune system can be summed up in many steps. First step involves activation of T cells to express CD154 and interaction of CD4 T cells with APC through CD40-CD154 interactions. This important CD40-CD154 interaction at this step regulates activation of CD4 T cells and priming and expansion of antigen specific T cells. This step is mediated by induction of costimulatory activity on APC, activation of APC to produce cytokines, such as IL-12 or both. In addition, polarization of the immune response toward Th2 could also result from CD40-CD154 interaction by tight regulation of IL-12 production. Thus, CD40-CD154 interaction in this step will be important for many T cell effector functions, such as help for B cells, activation of DC, activation of monocytes and macrophages to produce cytokines and to kill intracellular pathogens, and activation of autoreactive T cells to mount an autoimmune response. In the second step, the migration of activated T cells to secondary lymphoid organs as well as to target organs is affected by CD40-CD154 interactions by regulation of adhesion molecules on endothelia. In the final phase, CD40-CD154 interactions may be important at the effector stage of the immune response. For example, CD40-CD154 interaction could activate inflammatory cells to secrete cytokines at the site of inflammation, upregulate costimulatory activity on local APCs to amplify the response and finally induce the secretion of inflammatory mediators required for an inflammation. Since CD154 is secreted by platelets and soluble CD154 has been found to be the biologically active form in many situations, it is likely that CD154 may exert its effect elsewhere away from the site of inflammation. Finally, it is possible that soluble CD154, which is readily available and shown to activate many cell types of the innate immune system, could potentially serve as early regulator of the innate immune response either on its own or in concert with other mediators of the innate immune response. Data reviewed in this article strongly support all of the above-mentioned possibilities, and provide a satisfactory basis to explain the profound effects on the immune system seen by interrupting CD40-CD154 interactions by blocking antibodies or by CD154 mutation in mice and humans.

Potential of CD40-CD154 as a Therapeutic Target

Many costimulatory pathways important for B and T cell activation, particularly those belonging to TNF superfamily are potential targets for therapeutic intervention, and several molecules are in clinical trials now. Since CD40-CD154 interaction regulates diverse pathways of the immune system, therapeutic strategies designed to modulate this interaction will provide useful additional means to treat autoimmune, neurological and cardiovascular diseases, and to prevent graft rejection. Approaches may include the use of anti-CD40 or anti-CD154 antibodies, development of small antagonist molecules, and chimeric soluble proteins that can bind CD40 or CD154 as a single agent or in combination with other molecules like CTLA-4-Ig. Strategies could also be developed to target intracellular pathways of CD40 mediated signaling, which include TRAFs and other molecules associated in this pathway. These approaches are consistent with the myriad of preclinical data in humans and strongly point to CD40-CD154 interactions as a key target of therapy in human disease. Although these therapeutic approaches have been successful in experimental models, side effects must be weighed before applying to humans because CD40 and CD154 are expressed ubiquitously. Since initial human clinical trials are associated with increased thrombotic events in humans, we are faced with a challenge to design for safer therapeutics for CD40-CD154 target. In this respect, CD40/CD154 is likely to remain the subject of intense investigation over the forthcoming years.

References

1. Noelle RJ, Ledbetter JA, Aruffo A. CD40 and its ligand, an essential ligand-receptor pair for thymus-dependent B-cell activation. Immunol Today 1992; 13:431-433.
2. Foy TM, Shepherd DM, Durie FH et al. In vivo CD40-gp39 interactions are essential for thymus-dependent humoral immunity. II. Prolonged suppression of the humoral immune response by an antibody to the ligand for CD40, gp39. J Exp Med 1993; 178:1567-1575.
3. Ramesh N, Fuleihan R, Geha R. Molecular pathology of X-linked immunoglobulin deficiency with normal or elevated IgM (HIGMX-1). Immunol Rev 1994; 138:87-104.
4. Callard RE, Armitage RJ, Fanslow W et al. CD40 ligand and its role in X-linked hyper-IgM syndrome. Immunol Today 1993; 14:559-564.
5. Notarangelo LD, Duse M, Ugazio AG. Immunodeficiency with hyper-IgM (HIM). Immunodef Rev 1992; 3:101-122.
6. Xu J, Foy TM, Laman JD et al. Mice deficient for the CD40 ligand. Immunity 1994; 1:423-431.
7. Renshaw BR, Fanslow WC, Armitage RJ et al. Humoral immune responses in CD40 ligand-deficient mice. J Exp Med 1994; 180:1889-1900.
8. Kawabe T, Naka T, Yoshida K et al. The immune responses in CD40-deficient mice: Impaired immunoglobulin class switching and germinal center formation. Immunity 1994; 1:167-178.
9. Grewal IS, Flavell RA. CD40 and CD154 in cell-mediated immunity. Annu Rev Immunol 1998; 16:111-135.
10. van Kooten C, Banchereau J. CD40-CD40 ligand. J Leukoc Biol 2000; 67:2-17.

11. Buchner K, Henn V, Grafe M et al. CD40 ligand is selectively expressed on CD4+ T cells and platelets: Implications for CD40-CD40L signalling in atherosclerosis. J Pathol 2003; 201:288-95.
12. Grewal IS, Flavell RA. CD40-CD40L interactions in T cell activation. Immunol Rev 1996; 153:85-106.
13. Grewal IS, Xu J, Flavell RA. Impairment of antigen-specific T-cell priming in mice lacking CD40 ligand. Nature 1995; 378:617-620.
14. Grewal IS, Foellmer HG, Grewal KD et al. Requirement for CD40 ligand in costimulation induction, T cell activation, and experimental allergic encephalomyelitis. Science 1996; 273:1864-1867.
15. Yang Y, Wilson JM. CD40 ligand-dependent T cell activation: Requirement of B7-CD28 signaling through CD40. Science1996; 273:1862-1864.
16. Grewal IS, Flavell RA. A central role of CD40 ligand in the regulation of CD4 T cell responses. Immunol Today 1996; 17:410-414.
17. Clark LB, Foy TM, Noelle RJ. CD40 and its ligand. Adv Immunol 1996; 63:43-78.
18. Foy TM, Aruffo A, Bajorath J et al. Immune regulation by CD40 and its ligand gp39. Ann Rev Immunol 1996; 14:591-617.
19. Banchereau J, Bazan F, Blanchard D et al. The CD40 antigen and its ligand. Annu Rev Immunol 1994;. 12:881-922.
20. Dallman C, Johnson PW, Packham G. Differential regulation of cell survival by CD40. Apoptosis 2003; 8:45-53.
21. Van Kooten C, Banchereau J. CD40-CD40 ligand: A multifunctional receptor-ligand pair. Adv Immunol 1996; 61:1-77.
22. Armitage RJ, Fanslow WC, Strockbine L et al. Molecular and biological characterization of a murine ligand for CD40. Nature 1992; 357:80-82.
23. Karpusas M, Hsu YM, Wang JH et al. 2 A crystal structure of an extracellular fragment of human CD40 ligand. Structure 1995; 3:1426-1435.
24. Mach F, Schonbeck U, Sukhova GK et al. Functional CD40 ligand is expressed on human vascular endothelial cells, smooth muscle cells, and macrophages: Implications for CD40-CD40 ligand signaling in atherosclerosis. Proc Natl Acad Sci USA 1997; 94:1931-1936.
25. Carbone E, Ruggiero G, Terrazzano G et al. A New Mechanism of NK Cell Cytotoxicity Activation: The CD40-CD40 Ligand Interaction. J Exp Med 1997; 185:2053-2060.
26. Ho WY, Cooke MP, Goodnow CC et al. Resting and anergic B cells are defective in CD28-dependent costimulation of naive CD4+ T cells. J Exp Med 1994; 179:1539-1549.
27. Banchereau J, de Paoli P, Valle A et al. Long-term human B cell lines dependent on interleukin-4 and antibody to CD40. Science 1991; 251:70-72.
28. Grabstein KH, Maliszewski CR, Shanebeck K et al. The regulation of T cell-dependent antibody formation in vitro by CD40 ligand and IL-2. J Immunol 1993; 150:3141-147.
29. Gray D, Siepmann K, Wohlleben G. CD40 ligation in B cell activation, isotype switching and memory development. Semin Immunol 1994; 6:303-310.
30. Caux C, Massacrier C, Vanbervliet B et al. Activation of human dendritic cells through CD40 cross-linking. J Exp Med 1994; 180:1263-1272.
31. Peguet-Navarro J, Dalbiez-Gauthier C, Rattis FM et al. Schmitt D Functional expression of CD40 antigen on human epidermal Langerhans cells. J Immunol 1995; 155:4241-4247.
32. Cella M, Scheidegger D, Palmer-Lehman K et al. Ligation of CD40 on dendritic cells triggers production of high levels of interleukin-12 and enhances T cell stimulatory capacity: T-T help via APC activation. J Exp Med 1996; 184:747-752.

33. Koch F, Stanzl U, Jennewein P et al. High level IL-12 production by murine dendritic cells: Upregulation via MHC class II and CD40 molecules and downgerulation by IL-10. J Exp Med 1996; 184:741-746.
34. Noelle RJ, Foy T, Mackey M et al. The role of CD40 and its ligand in peripheral T-cell tolerance. In: Banchereau J, Dodet D, Schwartz R et al, eds. Immune Tolerance. Paris: Elseveir, 1996:135-140.
35. Borrow P, Tishon A, Lee S et al. CD40L-deficient mice show deficits in antiviral immunity and have an impaired memory CD8+ CTL response. J Exp Med 1996; 183:2129-4236.
36. Oxenius A, campbell KA, Maliszewski CR et al. CD40-CD40 Ligand interactions are critical in T-B Cooperation but not for other anti-viral CD4+ T cell functions. J Exp Med 1996; 183:2209-2218.
37. Yang Y, Su Q, Grewal IS et al. Transient subversion of CD40 ligand function diminishes immune responses to adenovirus vectors in mouse liver and lung tissues. J Virol 1996; 70:6370-6377.
38. Stout R, Suttles J. The many roles of CD40:CD40L interactions in cell-mediated inflammatory responses. Immunol Today 1996; 17:487-492.
39. Stout R, Suttles J, Xu J et al. Impaired T cell-mediated Macrophage activation in CD40 ligand-deficient mice. J Immunol 1996; 156:8-11.
40. Wagner Jr DH, Stout RD, Suttles J. Role of the CD40-CD40 ligand interaction in CD4+ T cell contact-dependent activation of monocyte interleukin-1 synthesis. Eur J Immunol 1994; 24:3148-154.
41. Suttles J, Evans M, Miller RW et al. T cell rescue of monocytes from apoptosis: Role of the CD40-CD40L interaction and requirement for CD40-mediated induction of protein tyrosine kinase activity. J Leuk Biol 1996; 60:651-657.
42. Tian L, Noelle RJ, Lawrence DA. Activated T cells enhance nitric oxide production by murine splenic macrophages through gp39 and LFA-1. Eur J Immunol 1995; 25:306-309.
43. Yellin MJ, Brett J, Baum D et al. Functional Interactions of T cells with endothelial cells: The role of CD40L-CD40-mediated signals. J Exp Med 1995; 182:1857-1864.
44. Karmann K, Hughes CC, Schechner J et al. CD40 on human endothelial cells: Inducibility by cytokines and functional regulation of adhesion molecule expression. Proc Natl Acad Sci USA 1995; 92:4342-4346.
45. Karmann K, Min W, Fanslow WC et al. Activation and homologous desensitization of human endothelial cells by CD40 ligand, tumor necrosis factor, and interleukin 1. J Exp Med 1996; 184:173-182.
46. Hollenbaugh D, Mischel-Petty N, Edwards CP et al. Expression of functional CD40 by vascular endothelial cells. J Exp Med 1995; 182:33-40.
47. Mantovani A, Bussolino F, Introna M. Cytokine regulation of endothelial cell function: From molecular level to the beside. Immunol Today 1997; 18:231-240.
48. Stout RD, Suttles J. T cell signaling of macrophage activation: Cell contact-dependent and cytokine signals. Austin: R.G. Landes 1995:185.
49. Springer TA. Traffic signals for lymphocyte recirculation and leukocyte emigration: The multistep paradigm. Cell 1994; 76:301-314.
50. Laman JD, de Smet BJ, Schoneveld A et al. CD40-CD40L interactions in atherosclerosis. Immunol Today 1997; 18:272-277.
51. Zhou X, Stemme S, Hansson GK. Evidence for a local immune response in atherosclerosis. CD4+ T cells infiltrate lesions of apolipoprotein-E-deficient mice. Am J Pathol 1996; 149:359-366.
52. Emeson EE, Shen ML, Bell CG et al. Inhibition of atherosclerosis in CD4 T-cell-ablated and nude (nu/nu) C57BL/6 hyperlipidemic mice. Am J Pathol 1996; 149:675-685.
53. Lichtman AH, Cybulsky M, Luscinskas FW. Immunology of atherosclerosis: The promise of mouse models. Am J Pathol 1996; 149:351-357.

54. Wick G, Schett G, Amberger A et al. Is atherosclerosis an immunologically mediated disease? Immunol Today 1995; 16:27-33.
55. Buchner K, Henn V, Grafe M et al. CD40 ligand is selectively expressed on CD4+ T cells and platelets: Implications for CD40-CD40L signalling in atherosclerosis. J Pathol 2003; 201:288-295.
56. Lutgens E, Gorelik L, Daemen MJ et al. Requirement for CD154 in the progression of atherosclerosis. Nat Med 1999; 5:1313-1316.
57. Lutgens E, Cleutjens KB, Heeneman S et al. Both early and delayed anti-CD40L antibody treatment induces a stable plaque phenotype. Proc Natl Acad Sci USA 2000; 97:7464-7469.
58. Mach F, Schonbeck U, Sukhova GK et al. Reduction of atherosclerosis in mice by inhibition of CD40 signalling. Nature 1998; 394:200-203.
59. Stüber E, Strober W, Neurath M. Blocking the CD40L-CD40 interaction in vivo specifically prevents the priming of T helper 1 cells through the inhibition of interleukin-12 secretion. J Exp Med 1996; 183:693-698.
60. Durie FH, Fava RA, Foy TM et al. Prevention of collagen-induced arthritis with an antibody to gp39, the ligand for CD40. Science 1993; 261:1328-1330.
61. Mohan C, Shi Y, Laman JD et al. Interaction between CD40 and its ligand gp39 in the development of murine lupus nephritis. J Immunol 1995; 154:1470-1480.
62. Gerritse K, Laman JD, Noelle RJ et al. CD40-CD40 ligand interactions in experimental allergic encephalomyelitis and multiple sclerosis. Proc Natl Acad Sci USA 1996; 93:2499-2504.
63. Griggs ND, Agersborg SS, Noelle RJ et al. The relative contribution of the CD28 and gp39 costimulatory pathways in the clonal expansion and pathogenic acquisition of self-reactive T cells. J Exp Med 1996; 183:801-810.
64. Koshy M, Berger D, Crow MK. Increased expression of CD40 ligand on systemic lupus erythematosus lymphocytes. J Clin Invest 1996; 98:826-837.
65. Yellin MJ, Thienel U. T cells in the pathogenesis of systemic lupus erythematosus: Potential roles of CD154-CD40 interactions and costimulatory molecules. Curr Rheumatol Rep 2000; 2:24-31.
66. Yellin MJ, D'Agati V, Parkinson G et al. Immunohistologic analysis of renal CD40 and CD40L expression in lupus nephritis and other glomerulonephritides. Arthritis Rheum 1997; 40:124-134.
67. Kalled SL, Cutler AH, Datta SK et al. Anti-CD40 ligand antibody treatment of SNF1 mice with established nephritis: Preservation of kidney function. J Immunol 1998; 160:2158-2165.
68. Howard LM, Miga AJ, Vanderlugt CL et al. Mechanisms of immunotherapeutic intervention by anti-CD40L (CD154) antibody in an animal model of multiple sclerosis. J Clin Invest 1999; 103:281-290.
69. Ferguson TA, Stuart PM, Herndon JM et al. Apoptosis, tolerance, and regulatory T cells—old wine, new wineskins. Immunol Rev 2003; 193:111-123.
70. Larsen CP, Pearson TC. The CD40 pathway in allograft rejection, acceptance, and tolerance. Curr Opin Immunol 1997; 9:641-647.
71. Larsen CP, Elwood ET, Alexander DZ et al. Long-term acceptance of skin and cardiac allografts after blocking CD40 and CD28 pathways. Nature 1996; 381:434-438.
72. Elwood ET, Larsen CP, Cho HR et al. Prolonged acceptance of concordant and discordant xenografts with combined CD40 and CD28 pathway blockade. Transplantation 1998; 65:1422-1428.
73. Kirk AD, Harlan DM, Armstrong NN et al. CTLA4-Ig and anti-CD40 ligand prevent renal allograft rejection in primates. Proc Natl Acad Sci USA 1997; 94:8789-8794.

74. Kirk AD, Burkly LC, Batty DS et al. Treatment with Humanized monoclonal antibody against CD154 prevents acute renal allograft rejection in nonhuman primates. Nat Med 1999; 5:686-693.
75. Brodeur SR, Angelini F, Bacharier LB et al. C4b-binding protein (C4BP) activates B cells through the CD40 receptor. Immunity 2003; 18:837-848.
76. Clark EA, Craxton A. A CD40 bridge between innate and adaptive immunity. Immunity 2003; 18:724-725.
77. Elzey BD, Tian J, Jensen RJ et al. Platelet-mediated modulation of adaptive immunity. A communication link between innate and adaptive immune compartments. Immunity 2003; 19:9-19.
78. Tan J, Town T, Mullan M. CD40-CD40L interaction in Alzheimer's disease. Curr Opin Pharmacol 2002; 2:445-451.
79. Tan J, Town T, Paris D et al. Microglial activation resulting from CD40-CD40L interaction after beta-amyloid stimulation. Science 1999; 286:2352-2355.
80. Tan J, Town T, Crawford F et al. Role of CD40 ligand in amyloidosis in transgenic Alzheimer's mice. Nat Neurosci 2002; 5:1288-1293.
81. Wiley JA, Harmsen AG. CD40 ligand is required for resolution of Pneumocystis carinii pneumonia in mice. J Immunol 1995; 155:3525-3529.
82. Cailliez JC, Seguy N, Denis CM et al. Pneumocystis carinii: An atypical fungal micro-organism. J Med Vet Mycol 1996; 34:227-239.
83. Lu P, Urban JF, Zhou XD et al. CD40-mediated stimulation contributes to lymphocyte proliferation, antibody production, eosinophilia, and mastocytosis during an in vivo type 2 response, but is not required for T cell IL-4 production. J Immunol 1996; 156:3327-3333.
84. Campbell KA, Ovendale PJ, Kennedy MK et al. CD40 Ligand is required for protective cell-mediated immunity to Leishmania major. Immunity 1996; 4:283-289.
85. Kamanaka M, Yu P, Yasui T et al. Protective role of CD40 in Leishmania major infection at two distinct phases of cell-mediated immunity. Immunity 1996; 4:275-281.
86. Soong L, Xu J, Grewal IS et al. Disruption of CD40-CD40 Ligand interactions results is an enhanced susceptibility to Leishmania amazonesis infection. Immunity 1996; 4:263-273.
87. Yang Y, Trinchieri G, Wilson JM. Recombinant IL-12 prevents formation of blocking IgA antibodies to recombinant adenovirus and allows repeated gene therapy to mouse lung. Nature Med 1995; 1:890-893.
88. Chehimi J, Starr SE, Frank I et al. Impaired interleukin 12 production in human immunodeficiency virus-infected patients. J Exp Med 1994; 179:1361-1366.
89. Chougnet C, Wynn TA, Clerici M et al. Molecular analysis of decreased interleukin-12 production in persons infected with human immunodeficiency virus. J Infect Dis 1996; 174:46-53.
90. Meyaard L, Hovenkamp E, Pakker N et al. Interleukin-12 (IL-12) production in whole blood cultures from human immunodeficiency virus-infected individuals studied in relation to IL-10 and prostaglandin E2 production. Blood 1997; 89:570-576.
91. Subauste CS, Wessendarp M, Sorensen RU et al. CD40-CD40 ligand interaction is central to cell-mediated immunity against Toxoplasma gondii: Patients with hyper IgM syndrome have a defective type 1 immune response that can be restored by soluble CD40 ligand trimer. J Immunol 1999; 162:6690-6700.
92. Subauste CS, Wessendarp M. Human dendritic cells discriminate between viable and killed Toxoplasma gondii tachyzoites: Dendritic cell activation after infection with viable parasites results in CD28 and CD40 ligand signaling that controls IL-12-dependent and -independent T cell production of IFN-gamma. J Immunol 2000; 165:1498-1505.

93. Seguin R, Kasper LH. Sensitized lymphocytes and CD40 ligation augment interleukin-12 production by human dendritic cells in response to Toxoplasma gondii. J Infect Dis 1999; 179:467-474.
94. Subauste CS. CD154 and type-1 cytokine response: From hyper IgM syndrome to human immunodeficiency virus infection. J Infect Dis 2002;185(Suppl 1):S83-89.
95. Subauste CS, Wessendarp M, Smulian AG et al. Role of CD40 ligand signaling in defective type 1 cytokine response in human immunodeficiency virus infection. J Infect Dis 2001; 183:1722-1731.
96. Connors M, Kovacs JA, Krevat S et al. HIV infection induces changes in CD4+ T-cell phenotype and depletions within the CD4+ T-cell repertoire that are not immediately restored by antiviral or immune-based therapies. Nat Med 1997; 3:533-540.

Chapter 2

Regulation of T Cell Immunity by OX40 and OX40L

Michael Croft,* Shahram Salek-Ardakani, Jianxun Song, Takanori So and Pratima Bansal-Pakala

Abstract

OX40 (CD134) and its binding partner OX40-ligand (OX40L) represent members of the TNFR and TNF superfamilies that appear to be crucial to many types of immune reaction mediated by T cells. Emerging data have now put these molecules at the forefront of the field of what has been termed T cell costimulation. Costimulation is defined as signals from membrane bound molecules that synergize with, or modify, signals provided when the T cell encounters its specific antigen. In large part, costimulation is essential for an efficient T cell response, whether it is protective or pathogenic, and without costimulatory interactions between membrane bound receptor-ligand pairs, a T cell is ineffective and may often succumb to death or become nonfunctional. OX40 is induced on the T cell surface a number of hours or days after recognition of antigen, and expression coincides with the appearance of OX40L on several cell types that can present antigen such as dendritic cells and B cells. Recent data show that OX40 can provide signals to a T cell to allow prolonged cell division after activation and to prevent excessive cell death. The OX40/OX40L interaction then controls the absolute number of pathogenic or protective effector T cells that are generated at the peak of the immune response and dictates the frequency of memory T cells that subsequently develop. This then has implications regarding strategies to suppress unwanted immune responses, and for vaccination to promote naturally weak immune responses. Reagents that interfere with the binding of OX40 to OX40L have been shown to inhibit T cell responses and pathogenic symptoms in a number of immune based diseases. Conversely reagents that augment OX40 signals have now shown therapeutic efficacy in models of cancer. This article will review the literature regarding these molecules and discuss their implications in T cell immunity.

*Corresponding Author: Michael Croft—La Jolla Institute for Allergy and Immunology, Division of Immunochemistry, San Diego, California 92121, USA. Email: mick@liai.org

TNF Superfamily, edited by Sanjay Khare. ©2007 Landes Bioscience

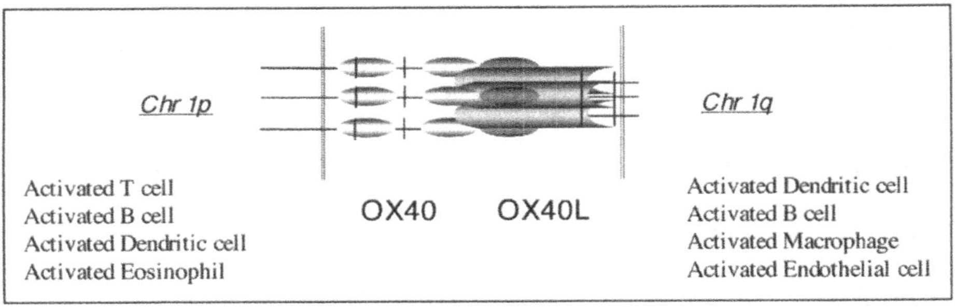

Figure 1. Schematic of OX40 and OX40L interaction. OX40, the type 1 transmembrane protein of the TNFR family is on the left with its characteristic cysteine-rich domains depicted. OX40L, the type II transmembrane protein of the TNF family, is on the right. It is likely that OX40L exists as a trimer and recruits 3 OX40 molecules into close proximity. The chromosomal location of the human genes are indicated, as well as the expression patterns on lymphoid cells.

Introduction to OX40 (CD134) and OX40-Ligand

OX40 is a 50 Kd glycosylated type 1 transmembrane protein.[1-4] The extracellular N-terminal portion of OX40 is 191 amino acids, and contains three cysteine-rich domains (CRDs) of approximately 40 amino acids, which are characteristic of the TNFR superfamily. The intracellular region consists of 36 amino acids. OX40L is a 34 Kd glycosylated type II transmembrane protein and as with other members of the TNF family is thought to be present on the surface of cells as a trimer.[5-7] The extracellular C-terminal domain of OX40L has a 133 amino acid long TNF homology domain (THD) that organizes into a characteristic "jelly roll" beta-sandwich structure. The intracellular region consists of 50 amino acids. Based on sequence similarity with other members of the TNF/TNFR family, the quaternary organization of the signaling unit of OX40 interacting with OX40L is likely to be three OX40 molecules bound to one trimeric OX40L complex (Fig. 1).[8,9]

The TNFR family of molecules can be divided into two groups based on the presence or absence of a cytoplasmic death domain that leads to apoptosis. OX40 lacks such a domain, and similar to other members of this family, extensive data from multiple systems now supports the idea that the main function of the OX40/OX40L interaction is to promote cell survival. This review will focus on recent studies of these molecules and discuss their role in immune function.

Expression Characteristics of OX40 and OX40L on T Cells and APC

OX40 was identified in 1987 with an antibody that reacted to activated rat CD4 T cells.[10] Although OX40 has been primarily visualized on CD4 T cells,[1,2] and the vast majority of studies to date have been directed towards this cell subset, CD8 T cells can also bear OX40 under certain conditions.[5,7] Moreover, OX40 has now been visualized on other diverse cell types including B cells, dendritic cells, and eosinophils (M. Croft et al, unpublished observations), although as yet the physiological significance of this expression is unknown.

OX40 is not expressed on resting T cells, but can be induced by peptide/MHC interaction with the T cell receptor (TCR) or reagents that cross-link the TCR/CD3 complex, and initially appears 12-24 hr after stimulation of naïve T cells. Peak expression is seen after 2-3 days and then OX40 is downregulated several days later, implying a delayed mode of action in primary immune responses. OX40 has been visualized in vivo in T cell zones of spleen or lymph nodes several days after immunization with protein antigen, directly coinciding with the peak of the primary T cell response.[11,12] In contrast, antigen-experienced memory/effector T cells can rapidly reexpress OX40 within 4 hr of reactivation, suggesting an earlier role in secondary immune responses of memory T cells when antigens are reencountered.[13]

OX40L was first identified on the surface of HTLV-infected leukemic T cells.[5] However, expression on nontransformed T cells appears to be rare and may depend on as yet undefined inflammatory conditions, or be restricted to specialized sites such as the gut.[14,15] The majority of OX40L is found on professional antigen-presenting cells (APC) such as dendritic cells, B cells and macrophages and as with OX40 on T cells, OX40L is induced many hours to days after APC receive an activating stimulus.[6,16-18] In the case of dendritic cells and B cells, Toll-like receptor signals induced by LPS can promote OX40L expression in addition to contributions from Ig signals and CD40 signals.[16-18] Additionally, OX40L has been visualized on activated endothelial cells in vitro, and in tissues from patients with lupus nephritis and inflammatory bowel disease, implying a role in promoting migration of OX40-expressing T cells into inflamed tissues, or providing signals to T cells to augment their activity in these peripheral sites.[19-21]

Function of OX40 on T Cells

Since its discovery, a number of in vitro studies using either receptor specific agonist antibodies, or cells transfected with OX40L, have shown that signals through OX40 can augment T cell responses, either in isolation, or in combination with signals from the Ig superfamily member, CD28, when it interacts with its ligands B7-1 or B7-2.[5,6,10,13,22-25] Although a number of activities have been described in vitro such as enhancing T cell proliferation, cytokine secretion, and cell survival, recent studies of knockout animals, with antagonist and agonist reagents in vivo, or receptor-deficient T cells, have aided tremendously in defining the physiological role of OX40 signals when they are received in the context of multiple other signals provided when an APC presents antigen to a T cell.

The initial reports of OX40- and OX40L-deficient mice demonstrated that CD4 T cell responses to the viruses LCMV and influenza, to several common protein antigens, and in contact hypersensitivity reactions, were markedly reduced.[12,18,26-28] Further in vivo data in OX40-deficient mice provided an indication of the role of OX40 in governing T cell immunity when a frequency analysis

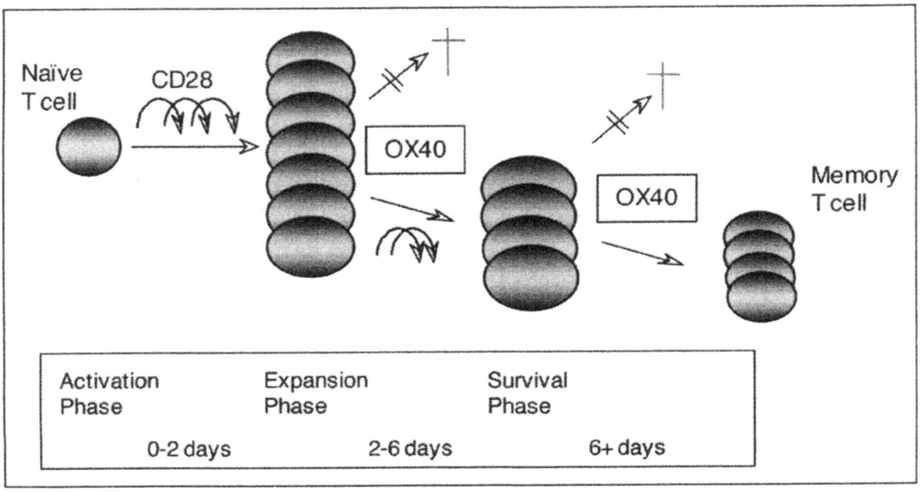

Figure 2. Temporal model of the role of OX40 in T cell responses. OX40 is induced 12 hrs or more after the encounter of a resting naïve T cell with an antigen-presenting cell (APC). OX40L is also induced on the APC, with peak expression and signaling through these molecules occurring 2-6 days after this initial activation phase. OX40 acts in a temporal manner after CD28 signals are initially provided to the T cell. OX40 can provide anti-apoptotic signals several days after a naïve T cell encounters antigen, and these signals allow continued turnover of cells and provide survival signals to prevent excessive T cell death at the peak of the primary response. Because of the strong anti-apoptotic action, OX40 signals are essential for allowing high numbers of memory T cells to develop which is the hallmark of effective T cell immunity.

of antigen-specific CD4 T cells generated after immunization showed dramatically reduced numbers late in the primary response, and additionally after 5 weeks when T cell memory was formed.[12] Corresponding data was produced when agonist anti-OX40 reagents, injected in vivo shortly after immunization, augmented the number of antigen-reactive CD4 cells that accumulated over time.[12,29] Along the same lines, transgenic expression of OX40L on dendritic cells in vivo led to greater numbers of primed CD4 cells[30] and blocking OX40L in another model reduced the accumulation of CD4 cells.[31]

Collectively, these studies have implied that a major role of OX40/OX40L interactions is to regulate the number of effector (protective or pathogenic) T cells that accumulate in primary immune responses, and consequently to promote a large number of memory T cells to subsequently develop (Fig. 2). More recent data obtained with antigen-specific TCR transgenic CD4 cells lacking OX40 have now provided greater insight into the mechanism of action.[32] These studies have directly demonstrated that OX40 signals contribute little to the initial response of a CD4 cell that occurs within 2-3 days of an encounter with antigen. OX40-deficient T cells become activated, secrete cytokines fairly normally, and go through a number of rounds of cell division. This directly contrasts with

CD28-deficient T cells in that this molecule is required for much of the early T cell response. The major phenotype seen in the absence of OX40 signals was reduced proliferation 4-5 days into the response, and a great defect in the ability to survive over the long-term. The lack of survival was shown to be due to apoptotic cell death and could be rescued by inhibitors of the caspase cascade.[32] OX40 expression is not dependent on CD28 signals, but several systems have shown that CD28 engagement can augment the level of OX40 expressed on a T cell.[31,32] As CD28 is constitutively expressed on a T cell, and then will provide signals prior to OX40 expression, this reinforces the concept that the two molecules most likely cooperate together in a sequential manner.

Therefore, in summary, the data at this point in time suggest a model whereby OX40 signals act in a temporal manner after CD28 signals, and enable effector T cells to survive and continue proliferating over an extended period of time, predominantly by transmitting anti-apoptotic signals that prevent excessive T cell death (Fig. 2). This ultimately results in greater numbers of T cells surviving the primary immune response and developing into memory T cells that can then respond in secondary immune reactions when antigen is reencountered at a later time.

Signals Transduced by OX40

How do OX40 signals regulate T cells and suppress cell death? As is the case with other members of the TNFR family, the intracellular tail of OX40 can bind several members of the TNFR-associated factor (TRAF) family of signaling molecules, in this case TRAF-2, -3 and -5.[33,34] The C-terminal region of TRAF2, which is conserved among the TRAF members, enables self-association into trimers, suggesting that OX40L binding brings three OX40 molecules into close proximity on the surface of a T cell and provides an opportunity for trimeric TRAF molecules to engage in multivalent interactions.[35] The interaction of the OX40 cytoplasmic tail and the C-terminal domain of TRAF molecules requires only a short stretch of conserved acidic amino acids, which contains the QEE motif.

NF-κB is most likely one of the central mediators of OX40 signals. After the recruitment of TRAF molecules to the cytoplasmic tail of activated OX40, TRAF-2 and -5 appear to play an important role in modulating an early step in activation of NF-κB by using their N-terminal RING and zinc finger domains. Studies using dominant negative forms of TRAF-2 and -5, which lack the N-terminal domains, demonstrated the critical contribution in OX40-induced NF-κB activation.[33,34] The introduction of TRAF3 together with the dominant negative mutants of TRAF2 or TRAF5 further reduced NF-κB activation,[36] suggesting that OX40 signals may be negatively modulated by TRAF3.

How OX40 provides survival signals remains incompletely understood (Fig. 3). Anti-apoptotic Bcl-2 family members (Bcl-2, Bcl-xL and Bfl-1) and pro-apoptotic Bcl-2-related proteins (Bim, Bad or Bid) have been identified that play key roles in regulating cell death in T cells. Recent data have shown that

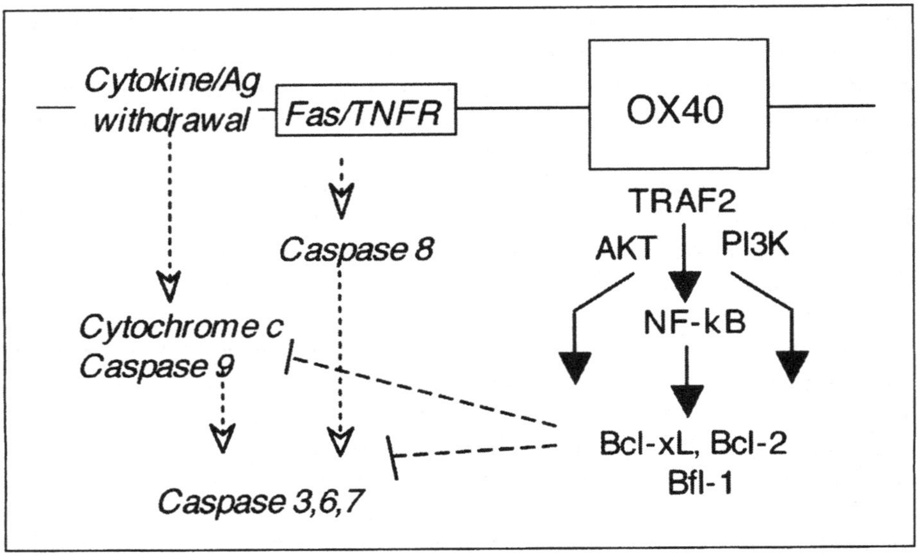

Figure 3. Signaling pathways induced when OX40 encounters OX40L. Functional data suggest that TRAF2 is responsible for many of the activities induced through OX40 by binding OX40L, although TRAF5 may also participate in this process. TRAF2 can associate with PI3K and activation of this molecule can lead to phosphorylation and activation of AKT. The downstream target of AKT is not clear, but NF-κB can also be activated by OX40 ligation, suggesting that this will also be important for the ultimate cellular effect. OX40 signals upregulate the expression of the anti-apoptotic members of the Bcl-2 family and block programmed cell death due to cytokine/antigen withdrawal. It is likely that AKT and NF-κB will mediate these activities but awaits a direct demonstration. It is not clear if Fas- or TNFRI-induced death can be inhibited by OX40 signals, but is a distinct possibility.

OX40-deficient CD4 cells cannot maintain high levels of Bcl-xL, Bcl-2, and Bfl-1 over the long-term following antigen stimulation.[32] Moreover, forced expression of Bcl-xL or Bcl-2 completely reversed the survival defect of these cells and conferred resistance to spontaneous apoptosis.[32] This data directly suggests that signals from OX40 positively affect molecules that inhibit the T cell from dying. Additionally, OX40 may also negatively affect the expression or activity of the pro-apoptotic molecules (J. Song and M. Croft, unpublished data) implying there may be two alternate coordinated modes of enhancing T cell survival. Whether TRAF2 or TRAF5, or activation of NF-κB activation, are required for these effects is presently unknown, but is obviously a distinct possibility.

Other recent studies have shown that a phosphatidylinostol 3-kinase (PI3K)-mediated signaling cascade mediates survival signals in multiple cell types. The serine/threonine protein kinase Bα (AKT/PKB) is a downstream target of

PI3K, and is also known to play an essential role in some forms of apopotic cell death. Recent data have also implicated PI3K and AKT in OX40 signaling. Ligation of OX40 induces PI3K recruitment and activation of AKT, and forced expression of active AKT in OX40-deficient T cells reverses their survival defect (J. Song and M. Croft, submitted for publication). How AKT inhibits T cell apoptosis is not clear. It has been shown to be capable of phosphorylating and inactivating Bad, directly up-regulating the expression of Bcl-xL and Bcl-2, and altering the function of transcription factors that can also lead to cellular apoptosis. It remains to be determined how PI3K and AKT modulate survival following OX40 ligation, but these molecules and NF-κB may be the principal intermediaries that regulate this activity of OX40 (Fig. 3).

Function and Signaling of OX40L on Accessory Cells

Although a lot of data exists on the role of OX40 on CD4 T cells, and obviously the interaction with OX40L is required for promoting OX40 signaling, it is not clear whether OX40L itself is essential for the response of the cell that bears it. Data gathered a number of years ago with reagents that cross-linked OX40L on B cells in vitro suggested that signals were transduced to allow a B cell to differentiate into a plasma cell secreting high levels of immunoglobulin.[16] This idea was initially supported when antibody production in vivo was suppressed by a polyclonal serum to OX40.[11] However, more recent data with OX40- and OX40L-deficient animals have now shown that these animals can mount relatively normal antibody responses,[26-28] questioning whether OX40L is required for a B cell response. A similar idea regarding APC activation was also put forward from other in vitro studies of dendritic cells that showed that these cells produced elevated levels of inflammatory cytokines such as IL-1 and IL-12 after OX40L was cross-linked on their surface.[17] As with the B cell studies, these data suggest there is the potential for OX40L to transduce signals to the APC at the time of encounter with an OX40-expressing T cell. However, more physiological data is needed from in vivo studies before it can be concluded that this is a major consequence of OX40/OX40L interaction.

As previously detailed, OX40L has also been seen on some activated endothelial cells[19] and the only study on signaling through OX40L has concentrated on this cell type. This data demonstrated that binding OX40L resulted in an increase in c-jun and c-fos mRNA, which is likely to be mediated by a cytoplasmic RPRF motif.[37] The endothelial cells were shown to upregulate production of a CC chemokine RANTES/CCL5 after OX40L engagement.[38] As this chemokine has been implicated in promoting migration of T cells into peripheral sites, these data suggest a possible link between OX40/OX40L and extravasation of T cells into inflamed tissues. However, again, it remains to be determined under more physiological conditions in vivo whether OX40L expression on endothelial cells is essential for a T cell response to develop at a site of inflammation.

Regulation of T Cell Tolerance and Cancer Immunity by OX40

It has been known for some time that recognition of antigen in a noninflammatory environment can lead to T cell tolerance,[39,40] and this process is characterized by defective survival of a large number of cells and hypo-responsiveness of those T cells that do survive. Because OX40 can be expressed on a T cell at low levels in the absence of inflammatory signals and CD28 signals are not essential, this suggests that OX40 may be a realistic target for providing signals to affect the tolerance process. This was directly shown in a recent study using agonist antibodies to OX40, where it was demonstrated that OX40 signals given within two days of encountering soluble antigen in vivo could prevent CD4 T cell deletion and stop T cells from entering a hypo-responsive state.[41] Similar data were also obtained assessing T cell deletion in response to a superantigen where agonist anti-OX40 also significantly enhanced the survival of CD4 T cells.[29] Moreover, additional data demonstrated that anti-OX40, given after T cell deletion and hyporesponsiveness had occurred, could target the small number of antigen-specific tolerant T cells and cause them to expand in numbers and to regain responsiveness.[41] The ability to reverse tolerance by providing costimulatory signals through OX40 indicates the possibility of a similar mechanism operating in vivo, if inadvertent OX40L expression were to occur, which would lead to autoimmunity. This was recently partly confirmed when transgenic mice were produced that constitutively expressed OX40L and were shown to spontaneously develop interstitial pneumonia, inflammatory bowel disease, and antibodies to double-stranded DNA.[42]

Therefore, if certain conditions are encountered that promote prolonged expression of both OX40 and OX40L, tolerant self-reactive T cells may gain normal function if they encounter their specific antigen and lead to development of late-onset autoimmune diseases. Blocking the interaction of OX40 and its ligand then represents a potential therapeutic target for limiting autoimmunity. On the other hand, treatment with agonist reagents to OX40 might be beneficial in situations where autoimmunity is desired.

Tumors can evade an immune response through an active tolerance mechanism by which T cells reactive with tumor self-peptides are deleted and/or made hyporesponsive, raising the possibility that OX40 targeted immunotherapy may be beneficial in augmenting anti-tumor immunity. OX40-expressing T cells have now been found at the sites of inflammation in patients with solid tumors,[43] directly on T lymphomas,[44-46] and within infiltrates of various types of tumors, including melanoma, head and neck cancer, breast cancer and colon cancer.[43,47,48] Treatment with reagents that bind to and signal through OX40 have recently been shown to delay tumor growth and enhance memory CD4 T cells reactive against tumor antigens.[49,50] For optimal systemic anti-tumor immunity, it is pertinent to develop strategies that would activate both CD4 and CD8 T cell responses against tumors. CD8 T cells upregulate OX40 upon activation, similar to CD4 T cells. Although the majority of data on OX40 have been directed towards CD4 T cells, agonistic anti-OX40 has also now been shown to enhance

CD8 T cell responses to antigen challenge in vivo (P. Bansal-Pakala and M. Croft, submitted for publication). OX40 signals can also reverse CD8 T cell tolerance (P. Bansal-Pakala and M. Croft, submitted), similar to CD4 cells.[41] This then holds great promise for anti-OX40 in tumor therapy, since by the time of therapeutic intervention, it is likely that T cells are already tolerized to existing tumors

Expression and Role of OX40 in T Cell-Mediated Disease

The expression of OX40 has now been detected on T cells at the site of inflammation in patients and rodents during clinical signs of a wide range of immunologically mediated diseases (Table 1) including: experimental allergic encephalomyelitis (EAE), the mouse model of MS,[51,52] graft-vs-host disease,[53-55] rheumatoid arthritis,[56,57] myasthenia gravis,[58] inflammatory bowel disease,[20] celiac disease, Crohn's disease, ulcerative colitis,[59] and inflammatory muscle disease.[60]

The expression patterns firstly suggest that OX40 may be a useful marker for identifying antigen-specific pathogenic T cells in a wide range of immune related diseases. Direct experimental evidence for this was first provided in EAE where the disease course was abrogated by administration of an anti-OX40 immunotoxin that directly killed CNS-infiltrating T cells.[61] More recently, it was shown that intravenous injection of immunoglobulin (IVIg) prevented the development of acute GVHD by decreasing the number of CD4+OX40+ donor alloreactive T cells.[62]

Secondly, expression of OX40 in sites of immune-mediated inflammation suggests that as well as OX40 providing a target for augmenting T cell function and enhancing immunity, the interaction of OX40 with OX40L is an attractive target for suppressing immune responses that may be detrimental to the host (Table 2). Studies of OX40-deficient mice showed that they exhibited reduced primary CD4 responses to the viruses LCMV and influenza, characterized by lower numbers of IFN-γ-secreting cells and fewer T cells infiltrating the lungs of infected animals.[26] OX40-deficient mice were also shown to have an impaired ability to generate Th2 immune responses and develop pulmonary lung inflammation and airway hyperreactivity in a murine model of asthma,[63] and this observation was confirmed in OX40L-deficient mice.[64] In other studies, OX40L-deficient mice were also found to be defective in primary contact hypersensitivity responses to oxazalone and DNBS.[27]

Several studies in experimental animal models have now not only stressed the importance of OX40 and OX40L for their manifestations, but shown that inhibiting this interaction can be useful therapeutically (Table 3). For example, anti-OX40L antibodies, or OX40-Ig fusion proteins that bind specifically to OX40L, can abrogate Th2 or Th1-induced pathologies in experimental leishmaniasis,[65] EAE,[66,67] acute graft-versus-host disease,[54,55] inflammatory bowel disease,[68,69] and collagen-induced arthritis.[56] These studies have highlighted the broad reaching control of T cell responses by OX40 and OX40L, particularly those mediated by CD4 cells, and promoted this interaction to the forefront of potential therapies aimed at dampening T cell driven immune diseases.

Table 1. Patterns of OX40 and OX40L expression in human and animal models of disease

Disease	Cell Type	OX40	OX40L	References
Human diseases:				
Cancers				
PTCL	Tumor cells (37%:148 cases)	+		Jones et al, 1999
ALCL	8/47-17%	+		
AIL	15/16-94%	+		
Angiocentric lymphoma	4/4-100%	+		
Large-cell lymphoma	10/21-48%	+		
Lymphoma with a prominent histiocytic component	6/7-86%	+		
Hodgkin's lymphoma	7/20-RS cells	+		
ATL (adult T-cell leukemia)	Leukemic cells: 15/17	+	–	Imura et al, 1997
Primary colon cancers	Lymphocytes	+		Petty et al, 2002
Primary breast tumors	Lymphocytes	+		Ramstad et al, 2000; Weinberg et al. 2000
Autoimmune disorders				
Proliferative lupus nephritis	Infiltrating leukocytes Tubular epithelium Endothelial cells	+ +	 +	Aten et al, 2000
Acute myocarditis	Infiltrating cells	+		Seko et al, 2002; Seko et al, 1999
Dilated cardiomyopathy	Cardiac monocytes		+	
RA	T cells	+		Giacomelli et al, 2001; Brugnoni et al, 1998; Yoshioka et al, 2000; Saijo et al, 2002
Polymyositis	Mononuclear cells	+		Tateyama et al, 2002
Granulomatous myopathy				
Myasthenia gravis	T cells	+		Onodera et al, 2000
Celiac disease	Gastrointestinal tissues	+		Stuber et al, 2000; Souza et al, 1999
Crohn's disease				
Ulcerative colitis				

continued on next page

Table 1. Continued

Disease	Cell Type	OX40	OX40L	References
Murine models of disease:				
Transplantation				
aGVHD	$CD4^+$ T cells	+		Tittle et al, 1997; Kotani et al, 2001
cGVHD	$CD8^+$ T cells	+		
Small bowel allografts		+	+	Tian et al, 2002
Autoimmune disorders				
EAE	T-cells	+		Weinberg et al, 1999;
	Macrophages		+	Nohara et al, 2001;
	Dendritic cells		+	Ndhlovu et al, 2001
	Endothelial cells		+	
IBD	Infiltrating leukocytes	+		Higgins et al, 1999; Malmstrom et al, 2001
	Activated DCs		+	
Infectious disease				
Leishmaniasis	$CD4^+$ T cells	+		Akiba et al, 2000
	Activated DCs		+	
Tumors	$CD4^+$ T cells	+		Kjaergaard et al, 2000
	$CD8^+$ T cells	+		

PTLC: peripheral T-cell lymphomas; ALCL: anaplastic large-cell lymphoma; AIL: angioimmunoblastic lymphoma RA: rheumatoid arthritis; aGVHD: acute graft versus host disease; cGVHD: chronic graft versus host disease; EAE: experimental allergic encephalomyelitis; IBD: inflammatory bowel diseases.

Summary

In conclusion, there is now a considerable body of literature that shows the importance of OX40 and OX40L in the generation of T cell immunity. Strong evidence has been presented that OX40 signaling to a T cell regulates expansion and cell division and is critical to the long-term survival of T cells. This not only impacts the magnitude of the primary immune response and its efficiency, but also directly affects the number of T cells that can go on to form the memory pool. Augmenting signals through OX40 have shown great promise in experimental models of tolerance and tumor immunity, and demonstrated that agonist reagents targeting OX40 may represent useful tools in the future as adjuvants for vaccination, and for treating ongoing immunological diseases that require a greater T cell response for protection. In addition, reagents that can inhibit endogenous OX40/OX40L interactions also show great promise as novel immunotherapeutic approaches for the treatment of autoimmune and allergic diseases that are characterized by exaggerated T cell responses.

Table 2. Major features of OX40 and OX40L deficient or transgenic mice

Disease	OX40-/- OX40L-/- OX40L-tg	Reduced Disease Severity	Enhanced Disease Susceptibility	Normal Response	References
Asthma	OX40-/-	YES			Jember et al, 2001
	OX40L-/-	YES			Arestides et al, 2002
CHS	OX40-/-	YES			Kopf et al, 1999
	OX40L-/-	YES			Sato et al, 2002
	OX40L-tg		YES		
EAE	OX40L-/-	YES			Ndhlovu et al, 2001
	OX40L-tg		YES		
Interstitial pmeumonia	OX40L-tg[a]		YES[b]		Murata et al, 2002
IBD	OX40L-tg[a]		YES[b]		Murata et al, 2002
Leishmaniasis	OX40-/-[c]			YES[c]	Pippig et al, 1999
Nippostrongylus brasiliensis	OX40-/-			YES	Pippig et al, 1999
Theilers murine encephalomyelitis	OX40-/-			YES	Pippig et al, 1999
LCMV	OX40-/-[d]			YES[e]	Kopf et al, 1999
VSV	OX40-/-			YES[e]	Kopf et al, 1999
	OX40L-/-				Chen et al, 1999
Influenza virus	OX40-/-			YES[f]	Kopf et al, 1999
DTH	OX40L-/-			YES	Chen et al, 1999

CHS: contact hypersentivity; EAE: experimental allergic encephalomyelitis; IBD: inflammatory bowel diseases; LCMV: lymphocytic choriomeningitis virus; VSV: vesicular stomatitis virus; DTH: delayed type hypersensitivity. [a] OX40L constitutively expressed on T-cells. These mice have elevated levels of serum Abs (IgM, IgG1, IgG2a, IgG2b, IgA, and IgE), cytokines IL-5 and IL-13, increased numbers of effector memory CD4 T-cells but not CD8 T-cells in the secondary lymphoid organs and absence of clonal T-cell deletion in response to superantigens. [b] OX40L-tg mice on C57BL/6 but not BALB/c background spontaneously develop interstitial pmeumonia and IBD. [c] No effect on viability or fertility; Normal primary and secondary lymphoid tissues; Leishmaniasis (129xC57BL/6). [d] Normal formation of GC, and Ab responses. [e] Reduced primary CD4 responses characterized by lower numbers of IFN-g-secreting cells and fewer T-cells infiltrating the lungs of infected animals. [f] Reduced primary CD4 responses. [g] OX40L overexoressed under the control of DC promoter: Increased numbers of CD4+ T-cells within B-cell follicles; Enlarged germinal centers; Brocker et al, 1999; Walker et al, 1999;

Table 3. Consequence of OX40 inhibition or OX40 engagement in animal models of disease

Disease	OX40 Ligand Blocking mAb	OX40 Agonist mAb	Disease Severity	References
aGVHD	OX40L mAb; OX40:Ig		Reduced lethality and disease severity	Weinberg et al, 1999; 2000; Tsukada et al, 2000; Stuber et al, 1998
Small bowel allograft	OX40L mAb		Reduced signs of disease	Tian et al, 2002
			Blocked development of colitis	Malmstrom et al, 2001;
IBD	OX40:Ig			Higgins et al, 1999
EAE	OX40L mAb; OX40:Ig		Reduced signs of disease	Stuber et al, 1998; Weinberg et al, 1999; Chitnis et al, 2001; Nohara et al, 2001
Leishmaniasis (susceptible BALB/c mice)	OX40L mAb		Inhibits development of disease	Akiba et al, 2000
Collagen induced arthritis	OX40L mAb		Inhibits development of disease	Yoshioka et al, 2000 Giacomelli et al, 2001
Response to sAg or SuperAg		α-OX40 mAb	Enhanced T-cell survival by inhibiting peripheral deletion	Maxwell et al, 2000
		α-OX40 mAb	Increase numbers of memory T-cells	Maxwell et al, 2000
Tolerance model		α-OX40 mAb	Breaks peripheral T-cell tolerance	Bansal-Pakala et al, 2001
Sarcoma, colon cancer		sOX40L: IgG	Enhanced anti-tumor responses and survival	Weinberg et al, 2000;
Glioma, melanoma				Kjaeraard et al, 2000
Lung, brain metastasis		α-OX40 mAb	Enhanced anti-tumor responses	Kjaeraard et al, 2001

aGVHD: acute graft versus host disease; IBD: inflammatory bowel diseases; EAE: experimental allergic encephalomyelitis; OX40L: OX40 ligand; mAb: monoclonal antibody; sAg: soluble antigens; SuperAg: superantigens. [a] OX40L mAb or OX40:Ig fusion protein: specifically inhibits OX40lOX40L interactions by binding to OX40 ligand. [b] α-OX40 mAb or sOX40L:Ig fusion protein: enhances OX40-mediated signaling by binding to OX40.

Future Considerations

Although much is known regarding the function of OX40 expressed on a T cell, there are many questions that remain unanswered not only regarding T cells but other cell types that have now been visualized to express OX40. For example, does OX40 contribute to the function of eosinophils or dendritic cells and if so, is it an analogous action to that on T cells. Additionally, even though a great amount of functional data has been obtained regarding the control of CD4 T cell responses, there is little data to date on the importance of OX40 signals to a CD8 T cell. Moreover, a number of CD4 T cell mediated responses such as those to helminth parasites, as well as several virally induced CD8 T cell responses, appear to be OX40 independent and the question is raised as to why OX40 is critical to certain T cell reactions but not others and whether there is a logical rationale for where and when OX40 plays a dominant role. Lastly, there is little information regarding the importance of OX40L signals to the cell type that expresses this molecule and whether it largely serves as an aggregation partner to induce OX40 signals, or whether there is a direct dialogue with the OX40L-expressing cell that is critical for the ultimate immune response. Answers to these questions will no doubt be forthcoming in the next few years and should provide a great framework for determining the therapeutic applications of targeting OX40 or OX40L in either positive or negative ways.

References

1. Mallett S, Fossum S, Barclay AN. Characterization of the MRC OX40 antigen of activated CD4 positive T lymphocytes—a molecule related to nerve growth factor receptor. EMBO J 1990; 9:1063-8.
2. Calderhead DM, Buhlmann JE van et al. Cloning of mouse Ox40: A T cell activation marker that may mediate T-B cell interactions. J Immunol 1993; 151:5261-71.
3. Latza U, Durkop H, Schnittger S et al. The human OX40 homolog: cDNA structure, expression and chromosomal assignment of the ACT35 antigen. Eur J Immunol 1994; 24:677-83.
4. Birkeland ML, Copeland NG, Gilbert DJ et al. Gene structure and chromosomal localization of the mouse homologue of rat OX40 protein. Eur J Immunol 1995; 25:926-30.
5. Baum PR, Gayle RB, Ramsdell F et al. Molecular characterization of murine and human OX40/OX40 ligand systems: Identification of a human OX40 ligand as the HTLV-1-regulated protein gp34. EMBO J 1994; 13:3992-4001.
6. Godfrey WR, Fagnoni FF, Harara MA et al. Identification of a human OX-40 ligand, a costimulator of CD4+ T cells with homology to tumor necrosis factor. J Exp Med 1994; 180:757-62.
7. Al-Shamkhani A, Birkeland ML, Puklavec M et al. OX40 is differentially expressed on activated rat and mouse T cells and is the sole receptor for the OX40 ligand. Eur J Immunol 1996; 26:1695-9.
8. Banner DW, D'Arcy A, Janes W et al. Crystal structure of the soluble human 55 kd TNF receptor-human TNF beta complex: Implications for TNF receptor activation. Cell 1993; 73:431-45.
9. Al-Shamkhani A, Mallett S, Brown MH et al. Affinity and kinetics of the interaction between soluble trimeric OX40 ligand, a member of the tumor necrosis factor superfamily, and its receptor OX40 on activated T cells. J Biol Chem 1997; 272:5275-82.

10. Paterson DJ, Jefferies WA, Green JR et al. Antigens of activated rat T lymphocytes including a molecule of 50,000 Mr detected only on CD4 positive T blasts. Mol Immunol 1987; 24:1281-90.
11. Stuber E, Strober W. The T cell-B cell interaction via OX40-OX40L is necessary for the T cell-dependent humoral immune response. J Exp Med 1996; 183:979-89.
12. Gramaglia I, Jember A, Pippig SD et al. The OX40 costimulatory receptor determines the development of CD4 memory by regulating primary clonal expansion. J Immunol 2000; 165:3043-3050.
13. Gramaglia I, Weinberg AD, Lemon M et al. Ox-40 ligand: A potent costimulatory molecule for sustaining primary CD4 T cell responses. J Immunol 1998; 161:6510-7.
14. Takasawa N, Ishii N, Higashimura N et al. Expression of gp34 (OX40 ligand) and OX40 on human T cell clones. Jpn J Cancer Res 2001; 92:377-82.
15. Wang HC, Klein JR. Multiple levels of activation of murine CD8(+) intraepithelial lymphocytes defined by OX40 (CD134) expression: Effects on cell-mediated cytotoxicity, IFN-gamma, and IL-10 regulation. J Immunol 2001; 167:6717-23.
16. Stuber E, Neurath M, Calderhead D et al. Cross-linking of OX40 ligand, a member of the TNF/NGF cytokine family, induces proliferation and differentiation in murine splenic B cells. Immunity 1995; 2:507-21.
17. Ohshima Y, Tanaka Y, Tozawa H et al. Expression and function of OX40 ligand on human dendritic cells. J Immunol 1997; 159:3838-48.
18. Murata K, Ishii N, Takano H et al. Impairment of antigen-presenting cell function in mice lacking expression of OX40 ligand. J Exp Med 2000; 191:365-74.
19. Imura A, Hori T, Imada K et al. The human OX40/gp34 system directly mediates adhesion of activated T cells to vascular endothelial cells. J Exp Med 1996; 183:2185-95.
20. Souza HS, Elia CC, Spencer J et al. Expression of lymphocyte-endothelial receptor-ligand pairs, alpha4beta7/MAdCAM-1 and OX40/OX40 ligand in the colon and jejunum of patients with inflammatory bowel disease. Gut 1999; 45:856-63.
21. Aten J, Roos A, Claessen N et al. Strong and selective glomerular localization of CD134 ligand and TNF receptor-1 in proliferative lupus nephritis. J Am Soc Nephrol 2000; 11:1426-38.
22. Flynn S, Toellner KM, Raykundalia C et al. CD4 T cell cytokine differentiation: The B cell activation molecule, OX40 ligand, instructs CD4 T cells to express interleukin 4 and upregulates expression of the chemokine receptor, blr-1. J Exp Med 1998; 188:297-304.
23. Kaleeba JA, Offner H, Vandenbark AA et al. The OX-40 receptor provides a potent costimulatory signal capable of inducing encephalitogenicity in myelin-specific CD4+ T cells. Int Immunol 1998; 10:453-61.
24. Ohshima Y, Yang LP, Uchiyama T et al. OX40 costimulation enhances interleukin-4 (IL-4) expression at priming and promotes the differentiation of naive human CD4(+) T cells into high IL-4-producing effectors. Blood 1998; 92:3338-45.
25. Akiba H, Oshima H, Takeda K et al. CD28-independent costimulation of T cells by OX40 ligand and CD70 on activated B cells. J Immunol 1999; 162:7058-66.
26. Kopf M, Ruedl C, Schmitz N et al. OX40-deficient mice are defective in Th cell proliferation but are competent in generating B cell and CTL responses after virus infection. Immunity 1999; 11:699-708.
27. Chen AI, McAdam AJ, Buhlmann JE et al. Ox40-ligand has a critical costimulatory role in dendritic cell: T cell interactions. Immunity 1999; 11:689-98.
28. Pippig SD, Pena-Rossi C, Long J et al. Robust B cell immunity but impaired T cell proliferation in the absence of CD134 (Ox40). J Immunol 1999; 163:6520-6529.

29. Maxwell J, Weinberg AD, Prell RA et al. Danger and OX40 receptor signaling synergize to enhance memory T cell survival by inhibiting peripheral deletion. J Immunol 2000; 164:107-112.
30. Brocker T, Gulbranson-Judge A, Flynn S et al. CD4 T cell traffic control: In vivo evidence that ligation of OX40 on CD4 T cells by OX40-ligand expressed on dendritic cells leads to the accumulation of CD4 T cells in B follicles. Eur J Immunol 1999; 29:1610-6.
31. Walker LS, Gulbranson-Judge A, Flynn S et al. Compromised OX40 function in CD28-deficient mice is linked with failure to develop CXC chemokine receptor 5-positive CD4 cells and germinal centers. J Exp Med 1999; 190:1115-22.
32. Rogers PR, Song J, Gramaglia I et al. OX40 promotes Bcl-xL and Bcl-2 expression and is essential for long-term survival of CD4 T cells. Immunity 2001; 15:445-55.
33. Arch RH, Thompson CB. 4-1BB and Ox40 are members of a tumor necrosis factor (TNF)-nerve growth factor receptor subfamily that bind TNF receptor-associated factors and activate nuclear factor kappaB. Mol Cell Biol 1998; 18:558-65.
34. Kawamata S, Hori T, Imura A et al. Activation of OX40 signal transduction pathways leads to tumor necrosis factor receptor-associated factor (TRAF) 2- and TRAF5-mediated NF- kappaB activation. J Biol Chem 1998; 273:5808-14.
35. Ye H, Park YC, Kreishman M et al. The structural basis for the recognition of diverse receptor sequences by TRAF2. Mol Cell 1999; 4:321-30.
36. Takaori-Kondo A, Hori T, Fukunaga K et al. Both amino- and carboxyl-terminal domains of TRAF3 negatively regulate NF-kappaB activation induced by OX40 signaling. Biochem Biophys Res Commun 2000; 272:856-63.
37. Matsumura Y, Hori T, Kawamata S et al. Intracellular signaling of gp34, the OX40 ligand: Induction of c-jun and c-fos mRNA expression through gp34 upon binding of its receptor, OX40. J Immunol 1999; 163:3007-11.
38. Kotani A, Hori T, Matsumura Y et al. Signaling of gp34 (OX40 ligand) induces vascular endothelial cells to produce a CC chemokine RANTES/CCL5. Immunol Lett 2002; 84:1.
39. Schwartz RH. Acquisition of immunologic self-tolerance. Cell 1989; 57:1073-81.
40. Schwartz RH. T cell clonal anergy. Curr Opin Immunol 1997; 9:351-7.
41. Bansal-Pakala P, Gebre-Hiwot Jember A, Croft M. Signaling through OX40 (CD134) breaks peripheral T-cell tolerance. Nat Med 2001; 7:907-12.
42. Murata K, Nose M, Ndhlovu LC et al. Constitutive OX40/OX40 ligand interaction induces autoimmune-like diseases. J Immunol 2002; 169:4628-36.
43. Vetto JT, Lum S, Morris A et al. Presence of the T-cell activation marker OX-40 on tumor infiltrating lymphocytes and draining lymph node cells from patients with melanoma and head and neck cancers. Am J Surg 1997; 174:258-65.
44. Durkop H, Latza U, Himmelreich P et al. Expression of the human OX40 (hOX40) antigen in normal and neoplastic tissues. Br J Haematol 1995; 91:927-31.
45. Imura A, Hori T, Imada K et al. OX40 expressed on fresh leukemic cells from adult T-cell leukemia patients mediates cell adhesion to vascular endothelial cells: Implication for the possible involvement of OX40 in leukemic cell infiltration. Blood 1997; 89:2951-8.
46. Jones D, Fletcher CD, Pulford K et al. The T-cell activation markers CD30 and OX40/CD134 are expressed in nonoverlapping subsets of peripheral T-cell lymphoma. Blood 1999; 93:3487-93.
47. Ramstad T, Lawnicki L, Vetto J et al. Immunohistochemical analysis of primary breast tumors and tumor- draining lymph nodes by means of the T-cell costimulatory molecule OX- 40. Am J Surg 2000; 179:400-6.

48. Petty JK, He K, Corless CL et al. Survival in human colorectal cancer correlates with expression of the T- cell costimulatory molecule OX-40 (CD134). Am J Surg 2002; 183:512-8.
49. Weinberg AD, Rivera MM, Prell R et al. Engagement of the OX-40 receptor in vivo enhances antitumor immunity. J Immunol 2000; 164:2160-9.
50. Kjaergaard J, Tanaka J, Kim JA et al. Therapeutic efficacy of OX-40 receptor antibody depends on tumor immunogenicity and anatomic site of tumor growth. Cancer Res 2000; 60:5514-21.
51. Weinberg AD, Wallin JJ, Jones RE et al. Target organ-specific up-regulation of the MRC OX-40 marker and selective production of Th1 lymphokine mRNA by encephalitogenic T helper cells isolated from the spinal cord of rats with experimental autoimmune encephalomyelitis. J Immunol 1994; 152:4712-21.
52. Weinberg AD, Lemon M, Jones AJ et al. OX-40 antibody enhances for autoantigen specific V beta 8.2+ T cells within the spinal cord of Lewis rats with autoimmune encephalomyelitis. J Neurosci Res 1996; 43:42-9.
53. Tittle TV, Weinberg AD, Steinkeler CN et al. Expression of the T-cell activation antigen, OX-40, identifies alloreactive T cells in acute graft-versus-host disease. Blood 1997; 89:4652-8.
54. Stuber E, Von Freier A, Marinescu D et al. Involvement of OX40-OX40L interactions in the intestinal manifestations of the murine acute graft-versus-host disease. Gastroenterology 1998; 115:1205-15.
55. Tsukada N, Akiba H, Kobata T et al. Blockade of CD134 (OX40)-CD134L interaction ameliorates lethal acute graft-versus-host disease in a murine model of allogeneic bone marrow transplantation. Blood 2000; 95:2434-9.
56. Yoshioka T, Nakajima A, Akiba H et al. Contribution of OX40/OX40 ligand interaction to the pathogenesis of rheumatoid arthritis. Eur J Immunol 2000; 30:2815-23.
57. Saijo S, Asano M, Horai R et al. Suppression of autoimmune arthritis in interleukin-1-deficient mice in which T cell activation is impaired due to low levels of CD40 ligand and OX40 expression on T cells. Arthritis Rheum 2002; 46:533-44.
58. Onodera J, Nagata T, Fujihara K et al. Expression of OX40 and OX40 ligand (gp34) in the normal and myasthenic thymus [In Process Citation]. Acta Neurol Scand 2000; 102:236-43.
59. Stuber E, Buschenfeld A, Luttges J et al. The expression of OX40 in immunologically mediated diseases of the gastrointestinal tract (celiac disease, Crohn's disease, ulcerative colitis). Eur J Clin Invest 2000; 30:594-9.
60. Tateyama M, Fujihara K, Ishii N et al. Expression of OX40 in muscles of polymyositis and granulomatous myopathy. J Neurol Sci 2002; 194:29-34.
61. Weinberg AD, Bourdette DN, Sullivan TJ et al. Selective depletion of myelin-reactive T cells with the anti-OX-40 antibody ameliorates autoimmune encephalomyelitis. Nat Med 1996; 2:183-9.
62. Caccavelli L, Field AC, Betin V et al. Normal IgG protects against acute graft-versus-host disease by targeting CD4(+)CD134(+) donor alloreactive T cells. Eur J Immunol 2001; 31:2781-90.
63. Jember AG, Zuberi R, Liu FT et al. Development of allergic inflammation in a murine model of asthma is dependent on the costimulatory receptor OX40. J Exp Med 2001; 193:387-392.
64. Arestides RS, He H, Westlake RM et al. Costimulatory molecule OX40L is critical for both Th1 and Th2 responses in allergic inflammation. Eur J Immunol 2002; 32:2874-80.
65. Akiba H, Miyahira Y, Atsuta M et al. Critical contribution of OX40 ligand to T helper cell type 2 differentiation in experimental leishmaniasis. J Exp Med 2000; 191:375-80.

66. Weinberg AD, Wegmann KW, Funatake C et al. Blocking OX-40/OX-40 ligand interaction in vitro and in vivo leads to decreased T cell function and amelioration of experimental allergic encephalomyelitis. J Immunol 1999; 162:1818-26.
67. Nohara C, Akiba H, Nakajima A et al. Amelioration of experimental autoimmune encephalomyelitis with anti- OX40 ligand monoclonal antibody: A critical role for OX40 ligand in migration, but not development, of pathogenic T cells. J Immunol 2001; 166:2108-15.
68. Higgins LM, McDonald SA, Whittle N et al. Regulation of T cell activation in vitro and in vivo by targeting the OX40-OX40 ligand interaction: Amelioration of ongoing inflammatory bowel disease with an OX40-IgG fusion protein, but not with an OX40 ligand-IgG fusion protein. J Immunol 1999; 162:486-93.
69. Malmstrom V, Shipton D, Singh B et al. CD134L expression on dendritic cells in the mesenteric lymph nodes drives colitis in T cell-restored SCID mice. J Immunol 2001; 166:6972-81.

CHAPTER 3

Signal Transduction in Osteoclast Biology: The OPG-RANKL-RANK Pathway

Ji Li*

Abstract

Skeletal homeostasis is maintained by a delicate balance between bone-resorbing osteoclasts and bone-building osteoblasts. Recently, three novel tumor necrosis factor (TNF) ligand and receptor family members have been identified as critical extracellular regulators of bone resorption: osteoprotegerin (OPG), receptor activator of nuclear factor NF-κB ligand (RANKL) or osteoprotegerin ligand (OPGL), and receptor activator of NF-κB (RANK). The subsequent characterization of the OPG-RANKL-RANK signal transduction pathway has elucidated the molecular mechanisms of osteoclast differentiation, activation, and survival, thus greatly expanded our basic understanding of osteoclast biology. In this new paradigm for the regulation of osteoclastogenesis and bone resorption, binding of RANKL to its transmembrane receptor RANK, expressed on the cell surface of hematopoietic osteoclast precursors and mature osteoclasts, initiates a signaling cascade that eventually leads to the differentiation and activation of osteoclasts. OPG, acting as a soluble decoy receptor, can bind to RANKL and uncouple the interaction between osteoblasts/stromal cells and osteoclast precursors, thereby inhibiting the osteoclast formation and maturation process. Both produced by osteoblasts and stromal cells, RANKL and OPG are regulated by various calciotropic hormones and pro-resorptive cytokines, and serve as the ultimate humoral mediator of bone resorption and calcium metabolism. In fact, it is the relative levels of RANKL and OPG expression that dictates the extent of bone resorption: excess RANKL increases bone resorption whereas excess OPG inhibit it. The biological importance of this pathway is underscored by the induction of extreme skeletal phenotypes, severe osteoporosis and osteopetrosis, via molecular genetic manipulation of these three extracellular signaling components in mice, as well as the identification of various genetic mutations in this pathway that cause several forms of rare human

*Ji Li—Amgen Inc., Department of Metabolic Disorders, MS 14-1-B, One Amgen Center Drive, Thousand Oaks, California 91320, USA. Email: jili@amgen.com

TNF Superfamily, edited by Sanjay Khare. ©2007 Landes Bioscience

genetic bone disorders. Perturbation of the OPG-RANKL-RANK signaling pathway has been implicated in the pathogenesis of many metabolic bone diseases, such as postmenopausal osteoporosis, bone loss associated with rheumatoid arthritis, tumor bone metastases, and humoral hypercalcemia of malignancy, while administration of OPG has been demonstrated to prevent or inhibit these osteolytic events in animal models that mimic these human disorders. More importantly, promising results from the first series of human clinical studies with recombinant OPG have further highlighted the therapeutic potential of targeting this signaling pathway for the treatment of osteolytic bone diseases such as osteoporosis and tumor bone metastases.

Manuscript

Bone remodeling and homeostasis is an essential function that regulates skeletal integrity throughout adult life in higher vertebrates and mammals. The maintenance of skeletal mass is controlled by the activities of specialized cells within the bone that have seemingly antagonistic activities: bone synthesis and bone resorption. Osteoblastic cells of mesenchymal origin synthesize and deposit bone matrix, and increase bone mass. Osteoclastic cells are large, multinucleated phagocytes of hematopoietic origin that resorb both mature and newly synthesized bone upon activation. Bone synthesis and resorption is a highly coordinated process, and requires a delicate balance between osteoclastic bone resorption and osteoblastic bone formation that is fine-tuned by a complex network of calciotropic and osteotropic hormones and cytokines, under physiological and pathological conditions.[1,2]

Until recently, the molecular mechanism that controls the cross talk (called coupling) between osteoblasts/stromal cells and osteoclast precursors, which lead to osteoclast formation, was poorly understood. However, recent advances have identified three new members of the tumor necrosis factor (TNF) ligand and receptor signaling system as essential regulatory components of osteoclastogenesis and bone resorption (Fig. 1). This novel cytokine system consists of: receptor activator of nuclear factor NF-κB ligand (RANKL) or osteoprotegerin ligand (OPGL), its cellular transmembrane receptor, receptor activator of NF-κB (RANK), and the soluble decoy receptor osteoprotegerin (OPG).

At Amgen, osteoprotegerin (OPG, protector of the bone) was discovered originally in a genomic effort opportunistically.[3] Seeking to unlock functions of novel secreted proteins, a transgenic mice platform was established that over-expresses these soluble factors systemically under the control of a liver specific apolipoprotein E promoter and associated enhancer. When transgenic mice were engineered to express a novel secreted TNF receptor (TNFR) family member hepatically, the mice were found to be osteopetrotic with denser, although normal shape bone. This unique soluble member of the TNFR family was then named osteoprotegerin based on this bone-protective biological activity. Independently, investigators at Snow Brand Milk Company in Japan identified an osteoclastogenic inhibitory factor (OCIF) from supernatant of a human fibroblast cell lines that inhibit

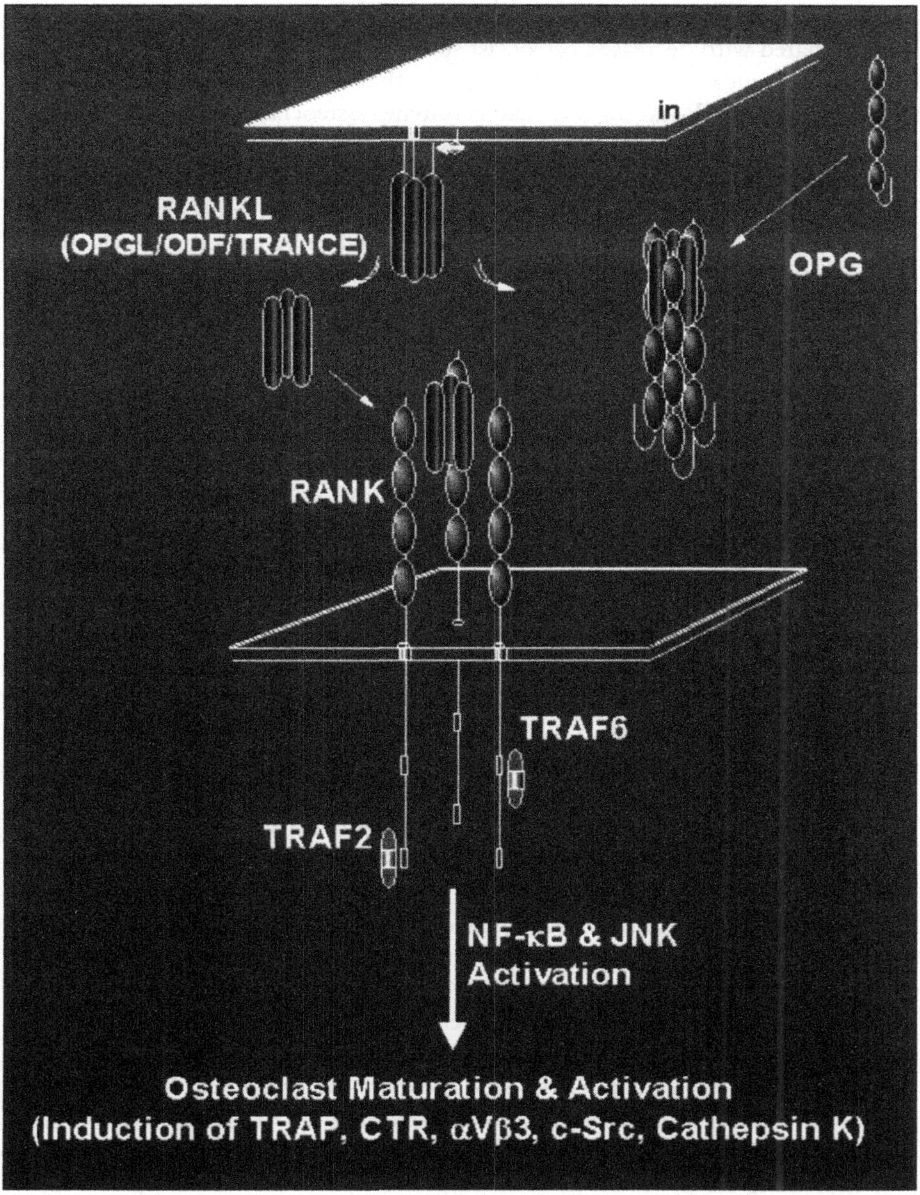

Figure 1. Signal transduction of OPG-RANKL-RANK pathway in osteoclast biology.

osteoclast formation in a osteoclast/stromal cell coculture system.[4] OCIF turned out to be encoded by the same gene as OPG.[5] Produced mainly by osteoblasts/stromal cells in bone, OPG has been shown to inhibit osteoclast differentiation, activation and survival in vitro.[3,5-7] In vivo, systemic administration of recombinant OPG protein in normal rodent resulted in rapidly increased bone mineral density that was associated with decreased osteoclast surface.[3]

The biological activity of OPG suggests that it might functions by neutralizing a critical TNF-like ligand that stimulates osteoclast development. Using OPG as a probe for expression cloning, RANKL/OPGL was isolated as a novel TNF family ligand to which OPG binds.[8,9] RANKL/OPGL turns out to be the long sought-after osteoclast differentiation factor (ODF)/stromal osteoclast forming activity (SOFA), which has been hypothesized to be the critical cell surface molecule that couples osteoblasts/stromal cells to osteoclasts.[1] RANKL, expressed primarily by osteoblasts/stromal cells and activated T lymphocytes, functions to stimulate osteoclast differentiation and activation, and to prolong osteoclast survival by inhibiting apoptosis.[7-10] In the presence of permissive concentrations of macrophage colony-stimulating factor (M-CSF) in vitro, RANKL is both necessary and sufficient for all phases of osteoclast development and thus, for bone resorption.[8,9] In vivo, direct administration of recombinant RANKL protein to normal mice resulted in dose-dependent, severe hypercalcemia, marked bone loss and profound osteoporosis, all due to an increased osteoclast activity. In contrast, cotreatment of these RANKL-treated mice with recombinant OPG completely blocked all these effects.[8]

The osteoclastogenic effects of RANKL imply the existence of a transmembrane signaling receptor on osteoclasts. Again, using RANKL as probe, RANK was isolated as a specific cellular receptor that transduces the actions of RANKL.[11] Binding of RANKL to RANK initiates a cascade of intracellular signaling events that eventually lead to the differentiation and activation of osteoclasts. This signal transduction process involves RANKL-dependent interaction with TNFR-associated factor (TRAF) family members, activation of the transcription factor NF-κB, and stimulation of the c-Jun N-terminal kinase (JNK) and Akt/PKB signaling pathway (Fig. 1).[11-15] Agonistic antibody directed against the extracellular domains of RANK mimicked the action of RANKL and promotes osteoclastogenesis, whereas inhibitory fragments of this RANK antibody or a soluble form of recombinant RANK blocks osteoclastogenesis.[11,16]

Some of the most compelling evidence supporting the critical role of the OPG-RANKL-RANK signaling system in osteoclast differentiation and activation comes from molecular genetic studies where transgenic and knockout mice for these molecules were generated and analyzed (Fig. 2). As mentioned above, the biological activity of OPG was initially realized via the analysis of transgenic mice over-expressing OPG.[3] These mice develop severe yet nonlethal osteopetrosis, as a result of markedly reduced osteoclast number with no concomitant reduction in macrophage number. In contrast, OPG knockout mice develop progressive and severe osteoporosis with high incidence of skeletal fractures.[17,18] In addition, mice with a disrupted RANKL gene show severe osteopetrosis, complete lack of osteoclasts as a result of an inability of osteoblasts/stromal cells from these mice to support osteoclastogenesis.[19] Similarly, RANK knockout mice also lack osteoclasts

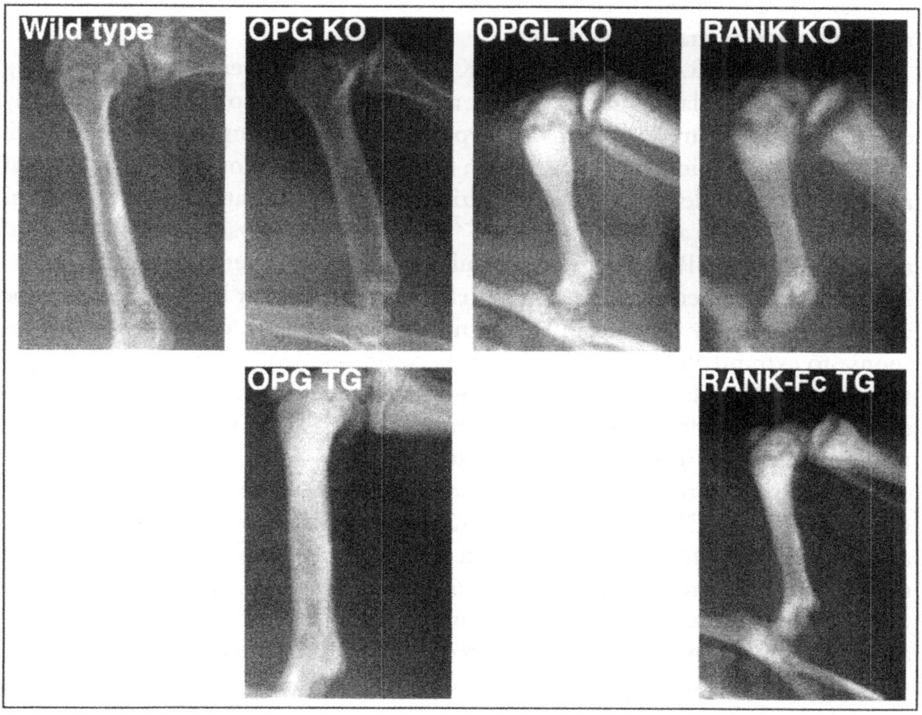

Figure 2. Molecular genetics of mouse osteoclast biology.

and have profound osteopetrosis.[20,21] However, unlike RANKL knockout mice, hematopoietic precursors isolated from RANK knockout mice are unable to form osteoclasts in vitro in the presence of RANKL and M-CSF, indicating an intrinsic defect in the osteoclast progenitors. Moreover, retroviral delivery of the RANK cDNA into hematopoietic precursors from RANK knockout mice successfully restored osteoclastogenesis, thus unequivocally confirmed RANK as the intrinsic hematopoietic cell surface determinant that controls osteoclastogenesis.[20] Overall, molecular genetic manipulations that inhibit the OPG-RANKL-RANK signal transduction pathway result in increases in bone mineral density and osteopetrosis, while manipulations that activate this signaling pathway lead to decreases in bone mineral density and osteoporosis (Fig. 2).

Another important function for the OPG-RANKL-RANK pathway has also been revealed during molecular genetic study of the same set of genetically engineered mouse models. In RANK or RANKL deficient knockout mice, the female animals fail to form lobulo-alveolar mammary gland structures during pregnancy and are not able to lactate, which leads to the death of newborn pups.[22] This surprising function has later been shown to be mediated by the IKK pathway downsteam of RANK.[23] Therefore, in addition to their critical role in regulating skeletal calcium release, RANKL and RANK are also essential for the development of lactating mammary gland.

While OPG, RANKL, and RANK were being discovered as key signaling molecules in osteoclast biology, RANKL and RANK have also been independently cloned from the immune system and hypothesized to play important role in T cell and dendritic cell biology.[24-26] Although neither the RANK or RANKL knockout mice showed any defect in T cell-dendritic cell communication, both knockout mice fail to develop any lymph nodes.[19-21] However, the physiological function of the OPG-RANKL-RANK system in adult immune system remains unclear. Most, if not all of the other immune system defects reported so far in these mice can be attributed either by the lack of lymph nodes development, or the secondary response to extramedullary hematopoiesis (a common phenotype in most osteopetrotic animals), while other potential immunological function of the pathway seems to be redundant with some other important TNF/TNFR molecules in the immune system, such as CD40/CD40 ligand.[19-21,27]

In addition to the overwhelming mouse genetic evidence supporting the essential role of the OPG-RANKL-RANK signaling pathway in osteoclast biology and bone resorption, several rare human genetic bone diseases have been mapped to defects in the same pathway due to enhanced signaling involving RANK and OPG. Three distinct tandem duplications in the 1st exon of RANK, presumably activating mutations, have been reported to cause autosomal dominant familial expansile osteolysis, early-onset Paget's disease of bone in Japan, and expansile skeletal hyperphosphatasia.[28,29] Idiopathic hyperphosphatasia (also known as Juvenile Paget's disease) is an autosomal recessive bone disease characterized by extremely rapid bone remodeling, osteopenia, fractures, and progressive skeletal deformity. It is reported recently that either homologous deletion or a single amino acid deletion of the human OPG gene can lead to this rare osteopathy.[30,31] All these human genetic evidence further demonstrated the importance of OPG-RANKL-RANK signaling pathway in human bone physiology and skeletal pathology.

In this newly emerged paradigm for the regulation of osteoclastogenesis and bone resorption (Fig. 1), binding of RANKL to its transmembrane receptor RANK, expressed on the cell surface of hematopoietic osteoclast precursors and mature osteoclasts, initiates a signaling cascade that eventually leads to the differentiation and activation of osteoclasts. OPG, acting as a soluble decoy receptor, can bind to RANKL and uncouple the interaction between osteoblasts/stromal cells and osteoclast precursors, thereby inhibiting the osteoclast formation and maturation process. Due to their critical function during osteoclast differentiation and activation, one can imagine that any molecule that regulates the relative abundance of RANKL and OPG, is likely to indirectly affect osteoclast biology. Most, if not all, calciotropic hormones and pro-resorptive cytokines have been shown to induce RANKL expression in various osteoblastic/stromal cell systems, and changes in the RANKL/OPG ratio has also been found to be associated with alteration in osteoclast differentiation and activation.[32,33] In fact, it is the relative levels of RANKL and OPG expression that dictates the extent of bone resorption: excess RANKL increases bone resorption whereas excess OPG inhibit it. The fact that recombinant OPG treatment and deletion of RANK receptor from mice can completely inhibit osteoclast formation and bone resorption induce by many of these calciotropic factors, strongly suggests that the OPG-RANKL-RANK signaling pathway is the ultimate common mediator of humoral signals regulating bone resorption and calcium metabolism.[20,34]

Imbalances between osteoclast and osteoblast activities often lead to inappropriately high bone resorption that is believed to be the cause for the majority of metabolic bone diseases, including postmenopausal osteoporosis. Abnormal changes in the expression and signaling of the OPG-RANKL-RANK pathway have been implicated in the pathogenesis of many of these osteolytic metabolic bone diseases.[32,33] Moreover, administration of recombinant OPG has been demonstrated to inhibit bone resorption in a variety of animal disease models, including ovariectomy-induced osteoporosis, bone loss associated with rheumatoid arthritis, experimental bone metastases, multiple myeloma, humoral hypercalcemia of malignancy and weightlessness.[3,34-40] Most importantly, the anti-resorptive therapeutic potential of OPG has been demonstrated recently in the first series of human clinical trials conducted with recombinant OPG protein.[41,42] In the first randomized, double-blind, placebo-controlled clinical study on postmenopausal women, a single subcutaneous (sc) dose of recombinant OPG causes rapid, profound and sustained inhibition of bone resorption dose dependently as indicated by the biochemical bone resorption marker profile.[41] In another recent phase I clinical study in patients with multiple myeloma or breast carcinoma related bone metastases, administration of a different form of recombinant OPG protein also led to a similar rapid, more sustained dose-dependant decrease in urinary bone resorption marker.[42] Both forms of recombinant OPG were safe and well tolerated.[41,42] Therefore, OPG could have wide application as a potent anti-resorptive therapeutic, with a distinct biological mechanism of action, for the treatment of various osteolytic metabolic bone diseases characterized by excessive bone resorption such as osteoporosis and tumor bone metastases. As one of the world's first therapeutic molecule coming directly out of the functional genomic approach being tested in human clinical study, the rapid progress of the OPG story from gene functioning and signaling pathway elucidation to human clinical testing is also a real testament of the power and potential of genomics in current biomedical research.

References

1. Rodan GA, Martin TJ. Role of osteoblasts in hormonal control of bone resorption—A hypothesis. Calcif Tissue Int 1981; 33(4):349-351.
2. Suda T, Takahashi N, Martin TJ. Modulation of osteoclast differentiation. Endocr Rev 1992; 13(1):66-80.
3. Simonet WS, Lacey DL, Dunstan CR et al. Osteoprotegerin: A novel secreted protein involved in the regulation of bone density. Cell 1997; 89(2):309-319.
4. Tsuda E, Goto M, Mochizuki S et al. Isolation of a novel cytokine from human fibroblasts that specifically inhibits osteoclastogenesis. Biochem Biophys Res Commun 1997; 234(1):137-142.
5. Yasuda H, Shima N, Nakagawa N et al. Identity of osteoclastogenesis inhibitory factor (OCIF) and osteoprotegerin (OPG): A mechanism by which OPG/OCIF inhibits osteoclastogenesis in vitro. Endocrinology 1998; 139(3):1329-1337.
6. Hakeda Y, Kobayashi Y, Yamaguchi K et al. Osteoclastogenesis inhibitory factor (OCIF) directly inhibits bone-resorbing activity of isolated mature osteoclasts. Biochem Biophys Res Commun 1998; 251(3):796-801.
7. Lacey DL, Tan HL, Lu J et al. Osteoprotegerin ligand modulates murine osteoclast survival in vitro and in vivo. Am J Pathol 2000; 157(2):435-448.
8. Lacey DL, Timms E, Tan HL et al. Osteoprotegerin ligand is a cytokine that regulates osteoclast differentiation and activation. Cell 1998; 93(2):165-176.
9. Yasuda H, Shima N, Nakagawa N et al. Osteoclast differentiation factor is a ligand for osteoprotegerin/osteoclastogenesis-inhibitory factor and is identical to TRANCE/RANKL. Proc Natl Acad Sci USA 1998; 95(7):3597-3602.
10. Burgess TL, Qian Y, Kaufman S et al. The ligand for osteoprotegerin (OPGL) directly activates mature osteoclasts. J Cell Biol 1999; 145(3):527-538.
11. Hsu H, Lacey DL, Dunstan CR et al. Tumor necrosis factor receptor family member RANK mediates osteoclast differentiation and activation induced by osteoprotegerin ligand. Proc Natl Acad Sci USA 1999; 96(7):3540-3545.
12. Galibert L, Tometsko ME, Anderson DM et al. The involvement of multiple tumor necrosis factor receptor (TNFR)- Associated factors in the signaling mechanisms of receptor activator of NF-kappaB, a member of the TNFR superfamily. J Biol Chem 1998; 273(51):34120-34127.
13. Wong BR, Josien R, Lee SY et al. The TRAF family of signal transducers mediates NF-kappaB activation by the TRANCE receptor. J Biol Chem 1998; 273(43):28355-28359.
14. Darnay BG, Haridas V, Ni J et al. Characterization of the intracellular domain of receptor activator of NF-kappaB (RANK). Interaction with tumor necrosis factor receptor-associated factors and activation of NF-kappab and c-Jun N-terminal kinase. J Biol Chem 1998; 273(32):20551-20555.
15. Wong BR, Besser D, Kim N et al. TRANCE, a TNF family member, activates Akt/PKB through a signaling complex involving TRAF6 and c-Src. Mol Cell 1999; 4(6):1041-1049.
16. Nakagawa N, Kinosaki M, Yamaguchi K et al. RANK is the essential signaling receptor for osteoclast differentiation factor in osteoclastogenesis. Biochem Biophys Res Commun 1998; 253(2):395-400.
17. Bucay N, Sarosi I, Dunstan CR et al. Osteoprotegerin-deficient mice develop early onset osteoporosis and arterial calcification. Genes Dev 1998; 12(9):1260-1268.
18. Mizuno A, Amizuka N, Irie K et al. Severe osteoporosis in mice lacking osteoclastogenesis inhibitory factor/osteoprotegerin. Biochem Biophys Res Commun 1998; 247(3):610-615.

19. Kong YY, Yoshida H, Sarosi I et al. OPGL is a key regulator of osteoclastogenesis, lymphocyte development and lymph-node organogenesis. Nature 1999; 397(6717):315-323.
20. Li J, Sarosi I, Yan XQ et al. RANK is the intrinsic hematopoietic cell surface receptor that controls osteoclastogenesis and regulation of bone mass and calcium metabolism. Proc Natl Acad Sci USA 2000; 97(4):1566-1571.
21. Dougall WC, Glaccum M, Charrier K et al. RANK is essential for osteoclast and lymph node development. Genes Dev 1999; 13(18):2412-2424.
22. Fata JE, Kong YY, Li J et al. The osteoclast differentiation factor osteoprotegerin-ligand is essential for mammary gland development. Cell 2000; 103(1):41-50.
23. Cao Y, Bonizzi G, Seagroves TN et al. IKKalpha provides an essential link between RANK signaling and cyclin D1 expression during mammary gland development. Cell 2001; 107(6):763-775.
24. Anderson DM, Maraskovsky E, Billingsley WL et al. A homologue of the TNF receptor and its ligand enhance T-cell growth and dendritic-cell function. Nature 1997; 390(6656):175-179.
25. Wong BR, Rho J, Arron J et al. TRANCE is a novel ligand of the tumor necrosis factor receptor family that activates c-Jun N-terminal kinase in T cells. J Biol Chem 1997; 272(40):25190-25194.
26. Wong BR, Josien R, Lee SY et al. TRANCE (tumor necrosis factor [TNF]-related activation-induced cytokine), a new TNF family member predominantly expressed in T cells, is a dendritic cell-specific survival factor. J Exp Med 1997; 186(12):2075-2080.
27. Bachmann MF, Wong BR, Josien R et al. TRANCE, a tumor necrosis factor family member critical for CD40 ligand- independent T helper cell activation. J Exp Med 1999; 189(7):1025-1031.
28. Hughes AE, Ralston SH, Marken J et al. Mutations in TNFRSF11A, affecting the signal peptide of RANK, cause familial expansile osteolysis. Nat Genet 2000; 24(1):45-48.
29. Whyte MP, Hughes AE. Expansile skeletal hyperphosphatasia is caused by a 15-base pair tandem duplication in TNFRSF11A encoding RANK and is allelic to familial expansile osteolysis. J Bone Miner Res 2002; 17(1):26-29.
30. Whyte MP, Obrecht SE, Finnegan PM et al. Osteoprotegerin deficiency and juvenile Paget's disease. N Engl J Med 2002; 347(3):175-184.
31. Cundy T, Hegde M, Naot D et al. A mutation in the gene TNFRSF11B encoding osteoprotegerin causes an idiopathic hyperphosphatasia phenotype. Hum Mol Genet 2002; 11(18):2119-2127.
32. Hofbauer LC, Khosla S, Dunstan CR et al. The roles of osteoprotegerin and osteoprotegerin ligand in the paracrine regulation of bone resorption. J Bone Miner Res 2000; 15(1):2-12.
33. Hofbauer LC, Heufelder AE. Role of receptor activator of nuclear factor-kappaB ligand and osteoprotegerin in bone cell biology. J Mol Med 2001; 79(5-6):243-253.
34. Morony S, Capparelli C, Lee R et al. A chimeric form of osteoprotegerin inhibits hypercalcemia and bone resorption induced by IL-1beta, TNF-alpha, PTH, PTHrP, and 1, 25(OH)2D3. J Bone Miner Res 1999; 14(9):1478-1485.
35. Kong YY, Feige U, Sarosi I et al. Activated T cells regulate bone loss and joint destruction in adjuvant arthritis through osteoprotegerin ligand. Nature 1999; 402(6759):304-309.
36. Capparelli C, Kostenuik PJ, Morony S et al. Osteoprotegerin prevents and reverses hypercalcemia in a murine model of humoral hypercalcemia of malignancy. Cancer Res 2000; 60(4):783-787.

37. Bateman TA, Dunstan CR, Ferguson VL et al. Osteoprotegerin mitigates tail suspension-induced osteopenia. Bone 2000; 26(5):443-449.
38. Morony S, Capparelli C, Sarosi I et al. Osteoprotegerin inhibits osteolysis and decreases skeletal tumor burden in syngeneic and nude mouse models of experimental bone metastasis. Cancer Res 2001; 61(11):4432-4436.
39. Pearse RN, Sordillo EM, Yaccoby S et al. Multiple myeloma disrupts the TRANCE/osteoprotegerin cytokine axis to trigger bone destruction and promote tumor progression. Proc Natl Acad Sci USA 2001; 98(20):11581-11586.
40. Croucher PI, Shipman CM, Lippitt J et al. Osteoprotegerin inhibits the development of osteolytic bone disease in multiple myeloma. Blood 2001; 98(13):3534-3540.
41. Bekker PJ, Holloway D, Nakanishi A et al. The effect of a single dose of osteoprotegerin in postmenopausal women. J Bone Miner Res 2001; 16(2):348-360.
42. Body JJ, Greipp P, Coleman RE et al. A Phase I study of AMGN-0007, a recombinant osteoprotegerin construct, in patients with multiple myeloma or breast carcinoma related bone metastases. Cancer 2003; 97(3 Suppl):887-892.

Chapter 4

Tumor Necrosis Factor (TNF) and Neurodegeneration

Rammohan V. Rao and Dale E. Bredesen*

Abstract

Cytokines are a family of growth factors that are secreted by the cells of the immune system. The family includes interleukins (IL), interferons (IFN), tumor necrosis factors (TNF), chemokines and other growth factors. Cytokines stimulate both the humoral and cellular immune responses as well as the activation of phagocytic cells and are generally associated with inflammation, immune activation, cell differentiation and cell death. They have diverse actions and are rapidly induced in response to tissue injury, infection or inflammation. Their role as mediators and inhibitors of diverse forms of neurodegeneration is increasingly recognized. Cytokines are induced in response to brain injury and can either induce, mediate, inhibit, or exacerbate cellular injury and repair. Several proinflammatory cytokines, notably tumor necrosis factor-α (TNF-α), have been shown to mediate diverse forms of experimental neurodegeneration, and both neurotoxic and neuroprotective actions have been reported. Here we review evidence for the contribution of cytokines to neurodegeneration, focusing primarily on tumor necrosis factor (TNF), which not only contributes to neuronal injury but may also exert protective effects. Since the mechanism of action of TNF and its interactions with various molecules in neurodegeneration is largely unknown, questions regarding these processes are of paramount importance to neurobiologists. Understanding the precise role of TNF in neurodegeneration is likely to have direct relevance in the search for potential treatments for neurodegenerative disease.

Apoptosis and Death Receptors

Apoptosis, a form of programmed cell death, is the most common physiological form of cell death. It plays a central role in normal embryonic development and homeostasis in adult tissues.[1] It occurs during embryonic development

*Corresponding Author: Dale E. Bredesen—Buck Institute for Age Research, 8001 Redwood Boulevard, Novato, California 94945, USA. Email: dbredesen@buckinstitute.org

TNF Superfamily, edited by Sanjay Khare. ©2007 Landes Bioscience

(e.g., in morphogenesis or synaptogenesis), tissue remodeling,[1] immune regulation[2] and tumor regression.[3] Apoptosis is essential for the normal development and maintenance of multicellular organisms as it clears individual cells without damaging the organism. Because the physiological role of apoptosis is crucial, aberration of this process can be catastrophic. Thus, uncontrolled apoptosis has been linked to nerve cell loss in conditions such as stroke, Alzheimer's and Parkinson's diseases,[4-8] whereas too little or failure of cells to initiate apoptosis contributes to cancer and autoimmune disease.[9]

Higher organisms have developed several mechanisms to clear cells rapidly and selectively by apoptosis. One such mechanism involves the interaction of surface receptors with their specific ligands.[10,11] Death receptors are cell surface receptors that transmit apoptotic signals initiated by specific death ligands. These receptors trigger cell death by acting as scaffolds for a class of enzymes called caspases. Caspases disassemble the cellular machinery causing an apoptotic demise of the cell. Death receptors are type 1 membrane proteins and belong to the tumor necrosis factor (TNF) receptor (TNFR) gene superfamily, defined by similar, cysteine-rich extracellular domains and in some but not all cases a homologous cytoplasmic sequence termed the "death domain" (DD).[12-14] These cytoplasmic death domains of TNFRs act as docking sites for signaling molecules. Death domains are about 60-80 amino acids long, are located in the receptor tail, and function as the adaptor domains that promote homotypic association and enable the death receptors to transmit cytotoxic signals.

The death receptors characterized to date include CD95 (also called Fas or Apo1) and TNFR1 (also called p55 or CD120a).[10,11,14,15] Additional death receptors are avian CAR1, death receptor 3 (DR3; also called Apo3, WSL-1, TRAMP, or LARD), DR4 and DR5 (also called Apo2, TRAIL-R2, TRICK 2, or KILLER).[10,11,14] The ligands that activate these receptors also display structural similarities, which are reflected by similar mechanisms of receptor recognition and activation. Almost all of the ligands are type II transmembrane proteins consisting of three identical subunits that activate their receptors by oligomerization.[16-19] For most members of the TNFR superfamily, specific ligands have been identified. CD95 ligand (CD95L or FasL) binds to CD95; TNF and lymphotoxin bind to TNFR1; Apo3 ligand (Apo3L, also called TWEAK) and Apo2 ligand (Apo2L, also called TRAIL) binds to DR4 and DR5.[10,11,14]

The low-affinity p75 neurotrophin receptor $p75^{NTR}$ is also a member of the tumor necrosis factor receptor superfamily that binds nerve growth factor (NGF) and other neurotrophins to regulate neuronal survival, differentiation and repair.[20-24] The extracellular domain of p75 has structural similarities to the tumor necrosis factor receptor and to the Fas antigen.[25,26] The p75-induced cell death has been shown to occur either as a ligand-dependent or independent phenomenon.[27-30]

The TNF System

TNF was first identified as a factor in endotoxin-primed mice that was capable of killing tumor cells in vitro and causing hemorrhagic necrosis in transplanted tumors. It plays a crucial role in various acute and chronic inflammatory disorders. Activated macrophages, lymphoid cells, NK cells, neutrophils, keratinocytes, and fibroblasts produce this cytokine in response to various stimuli.[10,11,14] TNF-α is expressed in all cell types in the CNS and has a wide range of biological effects including cytokine secretion, expression of adhesion molecules, induction of programmed cell death, antiviral activity, and activation of NF-κB.[10,11,14] TNF-α acts on two high affinity receptors, TNFR1 (p55) and TNFR2 (p75). Both receptors are expressed in the brain and signal through the NF-κB and MAPKs pathway. While most of the biological activities including antiviral activity, programmed cell death and activation of NF-κB are mediated by TNFR1, TNFR2 is mainly involved in stimulating the proliferation of T-lymphocytes.[31,32] The two forms of TNF, membrane bound and soluble forms bind to the two receptors with different affinities leading to diverse TNF responses.[33] Since both soluble and transmembrane forms of TNF-α can play critical roles in the pathogenesis of CNS inflammation and demyelination, blocking the activities of both these forms may be required to effectively neutralize the proinflammatory roles of this cytokine.[34]

TNF-α has been implicated in a variety of neurodegenerative diseases, including multiple sclerosis, stroke and ischemia, Alzheimer's disease and other age-associated neurodegenerative diseases[35-38] in which a pro-and anti-apoptotic role can be ascribed to TNF-α.[14,39] Induction of TNF-α mRNA expression has been observed as early as one hour after middle cerebral artery occlusion and exogenous TNF-α exacerbates focal ischemic injury.[40] While TNF-α has been shown to be proapoptotic, it also blocks necrosis in primary cortical neurons and triggers both apoptotic and necrotic cell death in PC12 cells.[41] TNF also mediates myelin and oligodendrocyte damage in vitro, exerts autocrine stimulatory effect on astrocytes, induces expression of MHC class 1 antigen on astrocytes and potentiates glutamate neurotoxicity in human fetal brain cell cultures.[42-44] $TNF^{(-/-)}$ mice display a normal developmental phenotype but are less sensitive than wild type mice to LPS-mediated toxicity and have impaired macrophage functions.[45,46] Overexpression of TNF-α by astrocytes or neurons triggers a spontaneous inflammatory and degenerative neurological disorder associated with chronic CNS inflammation and white matter degeneration.[34] However, overexpression of transmembrane TNF-α in astrocytes but not in neurons triggers neurological disorder in transgenic mice[47] suggesting that astrocytes and not neurons may form cellular contacts either with macrophages or endothelial cells that are critical for TNF to trigger the inflammatory response.

Although reported as being able to trigger neuronal death by either promoting apoptotic pathways[40,48,49] or suppressing survival signals,[50] TNF-α can also protect cultured embryonic rat hippocampal, septal, and cortical neurons against glucose deprivation-induced injury and excitatory amino acid toxicity.[51,52] Mice lacking the TNF receptors (TNFR-KO, both TNFR1 and TNFR2 deleted) demonstrated increased oxidative stress and damage to neurons caused by focal cerebral ischemia, as well as exacerbation of epileptic seizures, arguing that TNF serves a neuroprotective function.[52] This view was supported by work on Alzheimer's disease in which TNF-α was shown to have local neuroprotective effects by inducing anti-apoptotic pathways.[53,54] Thus, while high levels of TNF are injurious, low levels may be beneficial[55-57] and this may partly explain the dual (beneficial and injurious) role of TNF-α.

The TNF and TNFR1 have also evolved as mediators of cell death through their interactions with numerous death inducing molecules.[14] Binding of TNF to its receptor results in trimerization and the recruitment of TRADD, which in turn recruits FADD, TNFR-associated factor 2 (TRAF2) and the kinase-interacting protein RIP. FADD couples the TNFR1-TRADD complex to the activation of caspase-8, thereby initiating apoptosis (Fig. 1). Cells from FADD knockout mice are resistant to TNF-induced apoptosis, demonstrating an obligatory role of FADD in this response. TRAF2 is a member of the TRAF family of proteins, the latter of which associate with and transduce signals from TNF receptor family members.[10,11,14,15] TRAF2 also binds to cIAP1 and cIAP2 (cellular inhibitor of apoptosis-1 and -2),[58] which belong to a family of mammalian and viral proteins with anti-apoptotic activity. Besides FADD, TNFR1 can also engage an adapter called RAIDD or CRADD (Fig. 1). RAIDD binds through a death domain to the death domain of RIP and through a CARD motif to a similar sequence in the death effector caspase-2, thereby inducing apoptosis.[59,60] Thus, unlike the Fas system, which in almost all case signals cell death, the biological function of the TNF system is more complex and is reflected in the increased complexity of proteins that associate with the TNF-receptor signaling complex. Analysis of target genes of TNF-α under pathological conditions may provide clues about its role in various diseases including neurodegenerative disorders.

The $p75^{NTR}$ also contains a death domain and TRAF binding motifs near the carboxyterminus of the intracytoplasmic region.[24,61] TRAF proteins, TRAF1-TRAF6 have all been shown to interact with $p75,^{NTR}$ and at least TRAF2, 4, and 6 differentially modulate the ability of $p75^{NTR}$ to induce cell death and NF-κB activation.[62,63] While the death domains of TNFR and Fas interact with TRADD[64] and FADD/RIP[65,66] respectively, leading to the induction of apoptosis, the biochemical significance of the death domain of $p75^{NTR}$ is not yet clear.[26,67]

Neuroinflammation

The original notion that the brain represented an "immune-privileged" organ does not appear to be tenable in the light of recent reports demonstrating that the CNS can mount a well-defined inflammatory response to a variety of insults, as well as in neurodegenerative processes.[68] Inflammatory processes have been implicated in both acute and chronic neurodegenerative conditions.[68-71] Inflammatory

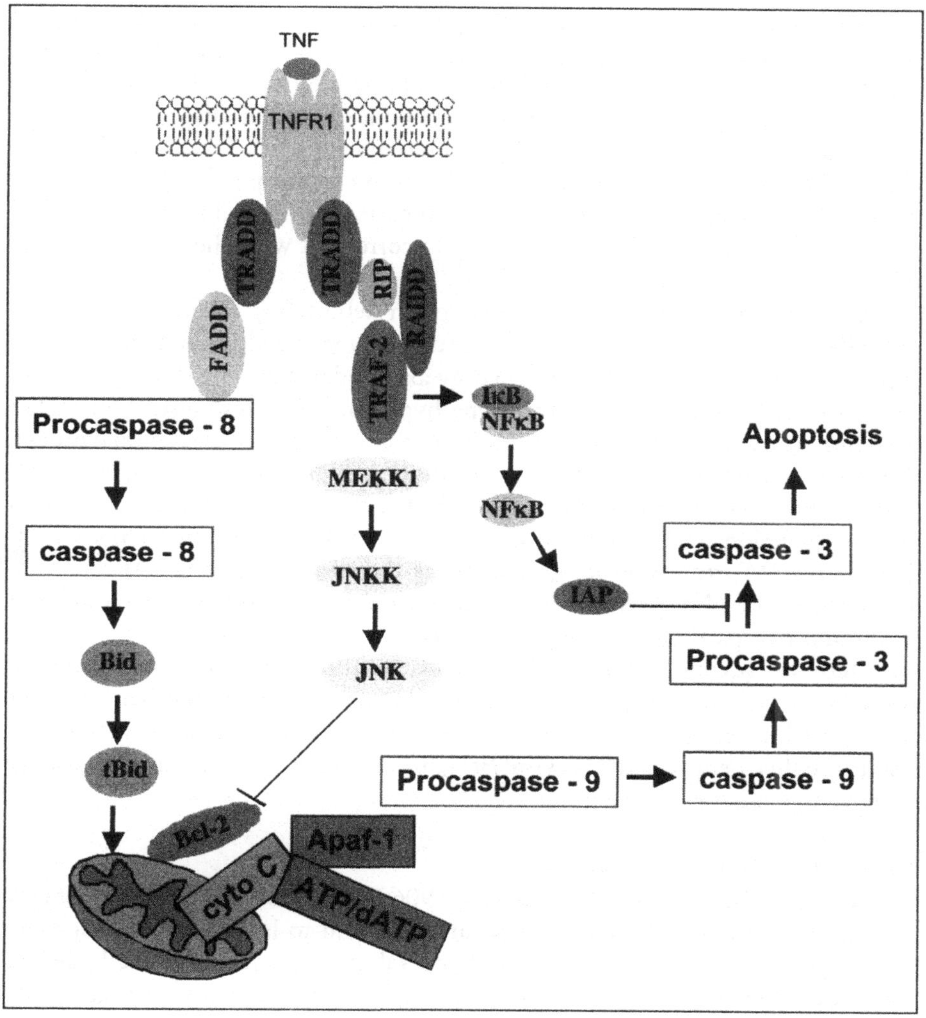

Figure 1. Signal transduction through the TNF TNFR system. Binding of TNF to its receptor results in the recruitment of TRADD, FADD, TRAF2 and RIP. FADD couples the TNFR1-TRADD complex to activation of caspase-8. Following cleavage by caspase-8, the C-terminus of Bid translocates (truncated Bid, tBid) from the cytosol to the mitochondria and causes the release of cytochrome c, thereby initiating apoptosis. Binding of TRAF2 and receptor-interacting protein (RIP) to the activated TNFR1 stimulates a protein kinase cascade pathway that includes MEKK1, JNKK, and JNK, leading to the activation of NF-κB and of JNK/AP-1, which regulate the expression of numerous immune and inflammatory response genes. TNF not only induces apoptosis by activating caspase-8, but can also inhibit apoptosis signaling via NF-κB, which induces the expression of IAP, an inhibitor of caspases 3, 7 and 9.

responses include the activation of astrocytes and microglia leading to increased production of classical inflammatory mediators, such as acute-phase proteins, eicosanoids, complement and cytokines including TNF.[39,72-77] Induction of the expression of pro and anti-inflammatory cytokines has been observed in focal or global ischemia, excitotoxicity, brain trauma and injury.[39] Several studies have also reported increased expression of cytokines in cerebrospinal fluid and in post mortem brain samples of stroke patients and patients with brain injury. The level of expression of cytokines in these samples correlated with the extent of tissue injury and with clinical outcome.[78,79]

Multiple sclerosis (MS) is a chronic, often debilitating autoimmune disease that affects the central nervous system and is characterized by localized areas of demyelination sometimes accompanied by axonal damage. The disease is associated with an inflammatory, delayed type hypersensitivity response. TNF-α has been implicated in the pathology of multiple sclerosis and a related animal model, experimental autoimmune encephalomyelitis (EAE). In brain lesions in MS, TNF positive cells were demonstrated and TNF was associated with astrocytes in all areas of the lesion, and with foamy macrophages in the center of the active lesion. TNF-α mRNA expression roughly parallels the clinical signs of EAE.[34,37,80-89] The presence of TNF in MS lesions suggests a significant role for cytokines and the immune response in disease progression.

Here we review the role of TNF in neurodegeneration, focusing on age-associated neurodegenerative diseases. Understanding the mechanisms that are involved in regulating TNF-α availability, sites and mechanism of action that result in either neuronal cell survival or neuronal death may lead to increasingly effective therapeutic interventions.

TNF and Alzheimer's Disease

First described by Alois Alzheimer in 1906, the disease that bears his name is the most common type of dementia occurring in mid-to-late life and is characterized by progressive loss of memory and general cognitive decline. The hallmark lesions of Alzheimer's disease (AD) include amyloid plaques containing aggregates of amyloid-β peptides (Aβ) derived from amyloid precursor protein (APP), neurofibrillary tangles (NFTs) (insoluble and highly phosphorylated forms of the microtubule-associated protein tau), and dystrophic neuritis.[90-93] To date, the cause and mechanism of progression of both familial and sporadic AD have not been fully elucidated. Recent studies have demonstrated a local inflammatory response in Alzheimer's disease (AD). The findings confirm a complex interaction between cytokines and amyloidogenesis in Alzheimer's disease, and indicate that astrocytes and microglia may be actively involved in cytokine-mediated events in AD. Circulating monocytes and macrophages, when recruited by chemokines produced by activated glial cells may add to the inflammatory destruction of the brain in Alzheimer's disease.[38,94-99] Amyloid-β peptides (Aβ) together with interferon-γ stimulate the activation of microglia, resulting in an increase in TNF-α release from the activated microglia. Concentrations of TNF-α and other cytokines have also been shown to be elevated in reactive glial cells that are in proximity to amyloid plaques. Thus, a combination of TNF-α, interferon-γ, transforming growth factor-beta (TGFβ), activated microglia and astrocytes and amyloid-β peptides may contribute to the neurotoxicity in AD.[38,54,94,95,99-103]

Recent reports also propose the involvement of the p75 neurotrophin receptor ($p75^{NTR}$) in the direct signaling of cell death by Aβ through its death domain. This signaling leads to the activation of caspases-8 and -3, the production of reactive oxygen intermediates and the induction of an oxidative stress. TNF-α and IL-1β, produced by Aβ-activated microglia, appear to potentiate the neurotoxic action of Aβ mediated by $p75^{NTR}$ signaling.[104-106] While serum levels of TNF have been reported to be elevated in AD,[35,107] other studies have found the levels to be unchanged or even decreased, thus leaving open the question of the role of TNF in AD pathogenesis.[96,108,109] It has also been argued that elevated levels of TNF in AD, rather than playing a role in the progression of the disease, may actually offer protection against disease progression.[54,110,111] Whatever the role of TNF in AD turns out to be, there is no doubt that local inflammatory responses occur in pathologically vulnerable regions of the AD brain. Damaged neurons and neurites, highly insoluble Aβ peptide deposits and neurofibrillary tangles provide stimuli for inflammation. By better understanding the inflammatory process in AD, it should be possible to develop anti-inflammatory approaches that, while unlikely to be curative, may slow the progression or delay the onset of this devastating disorder. Indeed, anti-inflammatory drugs such as those used in arthritis have been shown to delay or slow the progression of AD.[112-117]

TNF and Cerebral Ischemia

Ischemic cell death is initiated by decreased pH, decreased ATP levels, accumulation of free radicals, increased membrane depolarization and inhibition of oxidative phosphorylation. All these changes can trigger secondary changes in ion and chemical concentrations that ultimately activate the cell death process. Global, focal and hypoxia/ischemia are the three main laboratory models of ischemia. Global ischemia is most commonly produced by multiple vessel occlusions or cardiac arrest. A large part of the forebrain is quite uniformly affected. Focal ischemia can be induced by single vessel occlusion, cerebral hemorrhage, embolism or brain injury and results in a rapid neuronal death in a core region in the vicinity of the occluded artery. This is followed by a more delayed infarct in which neurons die several hours to days after the initial insult. Unilateral carotid occlusion in combination with hypoxia produces a hypoxic ischemia. Fifteen to thirty minutes following the insult, there is evidence of ischemic cell change and delayed cell death.

The expression of both pro- and anti-inflammatory cytokines is induced rapidly in all three models of ischemia. Within 30 minutes of focal ischemia, for example, TNF-α is elevated both in the circulation and in brain tissue itself. This increase in TNF-α has been shown to be damaging to the central nervous system.[118-120] Associated with this increase is the activation of NF-κB. Intracerebral injections of recombinant TNF-α markedly exacerbate ischemic tissue injury in vivo, as well as evoking cell death directly.[40,121] TNF-binding protein, a naturally occurring inhibitor of TNF, reduced damage caused by focal cerebral ischemia in mice. Similarly, blockade of TNF-α stimulation greatly enhanced the number of perfused microvessels at the end of MCA (middle cerebral artery) occlusion.[77,122,123] These findings indicate that endogenous TNF-α contributes, directly or indirectly, to neuronal injury. In contrast, however, other studies have supported a neuroprotective response of TNF-α. For example, mice lacking either the p55 TNF receptor (TNFR I) or both receptors (TNFR I and II) showed enhanced ischemic and excitotoxic injury compared with wild type animals and p75TNF receptor (TNFR II) deficient mice.[52,124] It is of course possible that TNF-α may enhance or inhibit neuronal injury depending on variables such as time course, extent of expression, receptor interactions, molecular complexes mediating its effect, etc. The mechanisms by which TNF-α protect neurons against ischemic and excitotoxic insults may involve induction of antioxidant pathways,[125] stabilization of intracellular calcium flux[126,127] and increasing the expression of antiinflammatory cytokines like IL-10 and TGF-α that play a role in repair mechanisms.[128-130]

TNF and Parkinson's Disease

Parkinson's disease (PD) is the second most common neurodegenerative disorder after Alzheimer's disease, and is characterized by the selective, progressive dysfunction and death of dopaminergic neurons (DA) in the substantia nigra. The characteristic symptoms of rigidity, bradykinesia, and resting tremor seen in PD are associated with loss of cells in the substantia nigra and depletion of dopamine in the striatum. Another important pathological feature is the presence of large intracytoplasmic inclusions called Lewy bodies that constitute degenerating ubiquitin-positive neuronal processes.[131,132] Although the exact mechanisms responsible for the DA neuronal cell loss are unclear, emerging evidence suggests the involvement of inflammatory events in neurodegenerative disorders including Parkinson's disease.[75,133-136]

Recent studies demonstrate that the loss of dopaminergic neurons is associated with a glial reaction and the overproduction of TNF-α in patients with PD.[135-138] Administration of MPTP, a dopaminergic neurotoxin that mimics some of the key features associated with PD, to wild-type mice resulted in an enhanced expression of TNF-α in striatum, preceding the loss of dopaminergic markers and reactive gliosis. MPTP-induced dopaminergic neurotoxicity was not observed in double transgenic mice (TNFR-DKO) carrying homozygous mutant alleles for both of the TNF receptors (TNFR1 and TNFR2), arguing that TNF-a is an obligatory component of dopaminergic neurodegeneration in this model.[137] TNF receptor (TNFR1) positive glial cells were found to be substantially higher in patients with PD than in controls. The percentage of FADD-immunoreactive dopaminergic (DA) neurons in the substantia nigra pars compacta of patients with PD was also higher compared with controls, suggesting that the TNF-receptor-ligand pair may participate in the degeneration of nigral DA neurons in PD.[75] Recent ultrastructural studies of dopaminergic neurons in patients with Parkinson's disease have shown that neurons die by apoptosis, preceded by the production of superoxide radicals in the mitochondria and the nuclear translocation of NF-κB.[133,139] TNF-α, observed in microglial cells in the substantia nigra of patients, may play a role, as might the transcription factor NF-κB.[140] The activities of caspase-8, caspase-9, caspase-1 and caspase-3 were also found to be significantly higher in the substantia nigra from parkinsonian patients than from control patients.[141-143] Thus TNF may participate in the degenerative processes that occur in Parkinson's disease. TNF-mediated changes in the levels of cytokines, neurotrophins, and caspases in the nigrostriatal regions of PD may be involved in apoptosis and degeneration of the nigrostriatal DA neurons.

TNF and Amyotrophic Lateral Sclerosis (Lou Gehrig's Disease)

The most common motor neuron disease in human adults is amyotrophic lateral sclerosis (ALS). The primary hallmark of ALS is the selective degeneration of motor neurons, the large nerve cells projecting from the motor cortex to the brainstem and spinal cord, and from the spinal cord to striated muscles. The loss of motor neurons leads to progressive atrophy of skeletal muscles, which initiates a progressive paralysis, typically in mid-life. In 90–95% of instances, there is no apparent genetic linkage, but in the remaining 5–10% of cases, the disease is inherited in a dominant manner. While the cause of motor neuron loss remains unknown, about 20% of patients with familial ALS (FALS) have mutations in the copper/zinc superoxide dismutase type 1 gene (*sod1*). However, the mechanism by which CuZnSOD (SOD1) mutant proteins provoke selective killing of motor neurons is not well understood. While several models have been proposed for the mechanism by which mutant CuZnSOD induces the motor neuron death characteristic of ALS, the three main mechanisms that have been advanced are: (a) the loss-of-function theory; (b) the oxidative stress hypothesis; and (c) the misfolded protein theory. The loss-of-function theory states that mutations in the *sod1* gene result in a defective form of CuZnSOD protein that has a reduced dismutase activity.[144-146] According to the second theory, mutant CuZnSOD has an altered oxidative enzymatic activity and promotes the ability of bound copper to engage in chemical reactions that produce hydroxyl radicals and/or reactive nitrogen species, resulting in the accumulation of oxidation products and nitrosylated proteins.[147-149] However, recent reports on (a) transgenic mouse models expressing mutant CuZnSOD protein,[146,150,151] (b) mice lacking the gene encoding the copper chaperone for CuZnSOD,[152] and (c) FALS model CuZnSOD mice lacking the gene encoding the copper chaperone for CuZnSOD[153] all argue against the first two theories. According to the third hypothesis, mutant proteins misfold and may form intracellular aggregates that, by an incompletely understood chemical mechanism, elicit toxicity, leading to organelle damage and ultimately motor neuron degeneration.[154-157]

In addition, an immunologic pathogenesis for ALS has also been proposed.[158-160] A dramatic neuroinflammatory reaction was observed in the brainstem and spinal cord of ALS cases and in mouse models of the disease.[158] Similarly, circulating levels of TNF-α, and the soluble forms of its receptors, were found to be increased in the blood of patients with ALS.[161] Peripherin, a neuronal intermediate filament protein associated with axonal spheroids in ALS was found to induce the degeneration of motor neurons and dorsal root ganglion neurons (DRG) when overexpressed in transgenic mice. The degeneration was mediated by the proinflammatory cytokine TNF.[162,163] Similar proinflammatory changes were also observed in spinal cords of transgenic mice expressing a CuZnSOD mutant. TNF-α expression was observed prior to the onset of motor deficits, and increased until the terminal stages of the disease.[164] cDNA microarray analysis of spinal cords from control and mice bearing the *sod1* gene with G93A mutation revealed an up-regulation of genes related to the inflammatory process and apoptosis, including TNF-α and caspase-1. The increased expression of the inflammation and apoptosis-related genes occurred at 11 weeks of age in the presymptomatic stage prior to motor neuron death, suggesting a mechanism of neurodegeneration associated with an inflammatory response.[165] Multiprobe ribonuclease protection assays were also performed to compare the expression of inflammatory cytokines and apoptosis-related genes in spinal cords of mice that ubiquitously express human CuZnSOD with the G93A mutation (G93A-SOD1). Caspases and death receptor complex components such as FADD and TNF-α were up-regulated at 120 days, a point at which the animals exhibit symptoms of motor neuron degeneration.[166] The results described above suggest that the mechanism of neurodegeneration in ALS may include an inflammatory response as an important component, with TNF-α and its receptors linking inflammation to apoptosis in ALS.

Summary and Future Directions

It is clear from many of the studies quoted above that cytokines, and especially TNF-α are induced after a number of CNS insults. It is also clear that TNF-α can act at very low concentrations within or outside the CNS. However, the precise role(s) of TNF-α in neurodegenerative disease pathogenesis remains incompletely understood. TNF-α may enhance or inhibit neuronal injury, depending on the duration of the insult, the extent of the expression of TNF and its receptors, and potentially on other variables. The mechanism of its action may depend on several factors including cell type, vasculature, frequency of insult, and other factors, which may be detrimental or beneficial. There is likely to be a significant interplay between TNF receptors and neuronal survival vs. death factors in neurodegeneration. Understanding the TNF modulated pathways is likely to facilitate the search for better therapeutic strategies for neurodegenerative diseases.

References

1. Milligan CE, Schwartz LM. Programmed cell death during animal development. Br Med Bull 1997; 53(3):570-590.
2. Ekert PG, Vaux DL. Apoptosis and the immune system. Br Med Bull 1997; 53(3):591-603.
3. Lyons SK, Clarke AR. Apoptosis and carcinogenesis. Br Med Bull 1997; 53(3):554-569.
4. Cheng Y, Deshmukh M, D'Costa A et al. Caspase inhibitor affords neuroprotection with delayed administration in a rat model of neonatal hypoxic-ischemic brain injury. J Clin Invest 1998; 101(9):1992-1999.
5. Namura S, Zhu J, Fink K et al. Activation and cleavage of caspase-3 in apoptosis induced by experimental cerebral ischemia. J Neurosci 1998; 18(10):3659-3668.
6. Jellinger KA. Cell death mechanisms in neurodegeneration. J Cell Mol Med 2001; 5(1):1-17.
7. Su JH, Zhao M, Anderson AJ et al. Activated caspase-3 expression in Alzheimer's and aged control brain: Correlation with Alzheimer pathology. Brain Res 2001; 898(2):350-357.
8. Cotman CW, Anderson AJ. A potential role for apoptosis in neurodegeneration and Alzheimer's disease. Mol Neurobiol 1995; 10(1):19-45.
9. Evan G, Littlewood T. A matter of life and cell death. Science 1998; 281(5381):1317-1322.
10. Ashkenazi A, Dixit VM. Death receptors: Signaling and modulation. Science 1998; 281(5381):1305-1308.
11. Schulze-Osthoff K, Ferrari D, Los M et al. Apoptosis signaling by death receptors. Eur J Biochem 1998; 254(3):439-459.
12. Tartaglia LA, Ayres TM, Wong GH et al. A novel domain within the 55 kd TNF receptor signals cell death. Cell 1993; 74(5):845-853.
13. Itoh N, Nagata S. A novel protein domain required for apoptosis. Mutational analysis of human Fas antigen. J Biol Chem 1993; 268(15):10932-10937.
14. Locksley RM, Killeen N, Lenardo MJ. The TNF and TNF receptor superfamilies: Integrating mammalian biology. Cell 2001; 104(4):487-501.
15. Nagata S. Apoptosis by death factor. Cell 1997; 88(3):355-365.
16. Jones EY, Stuart DI, Walker NP. Crystal structure of TNF. Immunol Ser 1992; 56:93-127.
17. Banner DW, D'Arcy A, Janes W et al. Crystal structure of the soluble human 55 kd TNF receptor-human TNF beta complex: Implications for TNF receptor activation. Cell 1993; 73(3):431-445.
18. Dhein J, Daniel PT, Trauth BC et al. Induction of apoptosis by monoclonal antibody anti-APO-1 class switch variants is dependent on cross-linking of APO-1 cell surface antigens. J Immunol 1992; 149(10):3166-3173.
19. Eck MJ, Ultsch M, Rinderknecht E et al. The structure of human lymphotoxin (tumor necrosis factor-beta) at 1.9- A resolution. J Biol Chem 1992; 267(4):2119-2122.
20. Coulson EJ, Reid K, Bartlett PF. Signaling of neuronal cell death by the p75NTR neurotrophin receptor. Mol Neurobiol 1999; 20(1):29-44.
21. Roux PP, Barker PA. Neurotrophin signaling through the p75 neurotrophin receptor. Prog Neurobiol 2002; 67(3):203-233.
22. Hempstead BL. The many faces of p75NTR. Curr Opin Neurobiol 2002; 12(3):260-267.
23. Ibanez CF. Neurotrophic factors: From structure-function studies to designing effective therapeutics. Trends Biotechnol 1995; 13(6):217-227.
24. Bibel M, Barde YA. Neurotrophins: Key regulators of cell fate and cell shape in the vertebrate nervous system. Genes Dev 2000; 14(23):2919-2937.

25. Meakin SO, Shooter EM. The nerve growth factor family of receptors. Trends Neurosci 1992; 15(9):323-331.
26. Bredesen DE. Keeping neurons alive: The molecular control of apoptosis (part I). The Neuroscientist 1996; 2:181-190.
27. Rabizadeh S, Oh J, Zhong LT et al. Induction of apoptosis by the low-affinity NGF receptor. Science 1993; 261(5119):345-348.
28. Rabizadeh S, Ye X, Wang JJ et al. Neurotrophin dependence mediated by p75NTR: Contrast between rescue by BDNF and NGF. Cell Death Differ 1999; 6(12):1222-1227.
29. Casaccia-Bonnefil P, Carter BD, Dobrowsky RT et al. Death of oligodendrocytes mediated by the interaction of nerve growth factor with its receptor p75. Nature 1996; 383(6602):716-719.
30. Frade JM, Rodriguez-Tebar A, Barde YA. Induction of cell death by endogenous nerve growth factor through its p75 receptor. Nature 1996; 383(6596):166-168.
31. Tartaglia LA, Goeddel DV, Reynolds C et al. Stimulation of human T-cell proliferation by specific activation of the 75-kDa tumor necrosis factor receptor. J Immunol 1993; 151:4637-4641.
32. Wong G, Kamb A, Tartaglia L et al. Possible protective mechanisms of tumor necrosis factors against oxidative stress. In: Scandalios JG, ed. Current communications in cell and molecular biology: Molecular biology of free radical scavenging systems. Cold Spring Harbor, New York: Cold Spring Harbor Laboratory Press, 1992:69-96.
33. Grell M, Douni E, Wajant H et al. The transmembrane form of tumor necrosis factor is the prime activating ligand of the 80 kDa tumor necrosis factor receptor. Cell 1995; 83(5):793-802.
34. Probert L, Akassoglou K, Kassiotis G et al. TNF-alpha transgenic and knockout models of CNS inflammation and degeneration. J Neuroimmunol 1997; 72(2):137-141.
35. Fillit H, Ding WH, Buee L et al. Elevated circulating tumor necrosis factor levels in Alzheimer's disease. Neurosci Lett 1991; 129(2):318-320.
36. Guan J, Skinner SJ, Beilharz EJ et al. The movement of IGF-I into the brain parenchyma after hypoxic-ischaemic injury. Neuroreport 1996; 7(2):632-636.
37. Hofman FM, Hinton DR, Johnson K et al. Tumor necrosis factor identified in multiple sclerosis brain. J Exp Med 1989; 170(2):607-612.
38. Blasko I, Marx F, Steiner E et al. TNFalpha plus IFNgamma induce the production of Alzheimer beta-amyloid peptides and decrease the secretion of APPs. Faseb J 1999; 13(1):63-68.
39. Allan SM, Rothwell NJ. Cytokines and acute neurodegeneration. Nat Rev Neurosci 2001; 2(10):734-744.
40. Barone FC, Arvin B, White RF et al. Tumor necrosis factor-alpha. A mediator of focal ischemic brain injury. Stroke 1997; 28(6):1233-1244.
41. Reimann-Philipp U, Ovase R, Weigel PH et al. Mechanisms of cell death in primary cortical neurons and PC12 cells. J Neurosci Res 2001; 64(6):654-660.
42. Selmaj KW, Raine CS. Tumor necrosis factor mediates myelin and oligodendrocyte damage in vitro. Ann Neurol 1988; 23(4):339-346.
43. Lavi E, Suzumura A, Murasko DM et al. Tumor necrosis factor induces expression of MHC class I antigens on mouse astrocytes. J Neuroimmunol 1988; 18(3):245-253.
44. Chao CC, Hu S. Tumor necrosis factor-alpha potentiates glutamate neurotoxicity in human fetal brain cell cultures. Dev Neurosci 1994; 16(3-4):172-179.
45. Pasparakis M, Alexopoulou L, Episkopou V et al. Immune and inflammatory responses in TNF alpha-deficient mice: A critical requirement for TNF alpha in the formation of primary B cell follicles, follicular dendritic cell networks and germinal centers, and in the maturation of the humoral immune response. J Exp Med 1996; 184(4):1397-1411.

46. Marino MW, Dunn A, Grail D et al. Characterization of tumor necrosis factor-deficient mice. Proc Natl Acad Sci USA 1997; 94(15):8093-8098.
47. Akassoglou K, Probert L, Kontogeorgos G et al. Astrocyte-specific but not neuron-specific transmembrane TNF triggers inflammation and degeneration in the central nervous system of transgenic mice. J Immunol 1997; 158(1):438-445.
48. Munoz-Fernandez MA, Fresno M. The role of tumour necrosis factor, interleukin 6, interferon-gamma and inducible nitric oxide synthase in the development and pathology of the nervous system. Prog Neurobiol 1998; 56(3):307-340.
49. Knoblach SM, Fan L, Faden AI. Early neuronal expression of tumor necrosis factor-alpha after experimental brain injury contributes to neurological impairment. J Neuroimmunol 1999; 95(1-2):115-125.
50. Venters HD, Tang Q, Liu Q et al. A new mechanism of neurodegeneration: A proinflammatory cytokine inhibits receptor signaling by a survival peptide. Proc Natl Acad Sci USA 1999; 96(17):9879-9884.
51. Cheng B, Christakos S, Mattson MP. Tumor necrosis factors protect neurons against metabolic-excitotoxic insults and promote maintenance of calcium homeostasis. Neuron 1994; 12:139-153.
52. Bruce AJ, Boling W, Kindy MS et al. Altered neuronal and microglial responses to excitotoxic and ischemic brain injury in mice lacking TNF receptors. Nat Med 1996; 2(7):788-794.
53. Tarkowski E, Blennow K, Wallin A et al. Intracerebral production of tumor necrosis factor-alpha, a local neuroprotective agent, in Alzheimer disease and vascular dementia. J Clin Immunol 1999; 19(4):223-230.
54. Mattson MP, Barger SW, Furukawa K et al. Cellular signaling roles of TGF beta, TNF alpha and beta APP in brain injury responses and Alzheimer's disease. Brain Res Brain Res Rev 1997; 23(1-2):47-61.
55. Shigeno T, Mima T, Takakura K et al. Amelioration of delayed neuronal death in the hippocampus by nerve growth factor. J Neurosci 1991; 11(9):2914-2919.
56. Mattson MP, Zhang Y, Bose S. Growth factors prevent mitochondrial dysfunction, loss of calcium homeostasis, and cell injury, but not ATP depletion in hippocampal neurons deprived of glucose. Exp Neurol 1993; 121(1):1-13.
57. Tracey KJ, Cerami A. Tumor necrosis factor, other cytokines and disease. Annu Rev Cell Biol 1993; 9:317-343.
58. Rothe M, Pan M-G, Henzel WJ et al. The TNFR2-TRAF signaling complex contains two novel proteins related to baculoviral inhibitor of apoptosis proteins. Cell 1995; 83:1243-1252.
59. Duan H, Dixit VM. RAIDD is a new 'death' adaptor molecule. Nature 1997; 385(6611):86-89.
60. Ahmad M, Srinivasula SM, Wang L et al. CRADD, a novel human apoptotic adaptor molecule for caspase-2, and FasL/tumor necrosis factor receptor-interacting protein RIP. Cancer Res 1997; 57(4):615-619.
61. Bibel M, Hoppe E, Barde YA. Biochemical and functional interactions between the neurotrophin receptors trk and p75NTR. EMBO J 1999; 18(3):616-622.
62. Khursigara G, Orlinick JR, Chao MV. Association of the p75 neurotrophin receptor with TRAF6. J Biol Chem 1999; 274(5):2597-2600.
63. Ye X, Mehlen P, Rabizadeh S et al. TRAF family proteins interact with the common neurotrophin receptor and modulate apoptosis induction. J Biol Chem 1999; 274(42):30202-30208.
64. Hsu H, Xiong J, Goeddel DV. The TNF receptor 1-associated protein TRADD signals cell death and NF- kappa B activation. Cell 1995; 81(4):495-504.

65. Chinnaiyan AM, O'Rourke K, Tewari M et al. FADD, a novel death domain-containing protein, interacts with the death domain of Fas and initiates apoptosis. Cell 1995; 81(4):505-512.
66. Boldin MP, Varfolomeev EE, Pancer Z et al. A novel protein that interacts with the death domain of Fas/APO1 contains a sequence motif related to the death domain. J Biol Chem 1995; 270(14):7795-7798.
67. Dobrowsky RT, Werner MH, Castellino AM et al. Activation of the sphingomyelin cycle through the low-affinity neurotrophin receptor. Science 1994; 265(5178):1596-1599.
68. Weiner HL, Selkoe DJ. Inflammation and therapeutic vaccination in CNS diseases. Nature 2002; 420(6917):879-884.
69. Perry VH, Bell MD, Brown HC et al. Inflammation in the nervous system. Curr Opin Neurobiol 1995; 5(5):636-641.
70. McGeer PL, McGeer EG. The inflammatory response system of brain: Implications for therapy of Alzheimer and other neurodegenerative diseases. Brain Res Brain Res Rev 1995; 21(2):195-218.
71. Floyd RA, Hensley K, Bing G. Evidence for enhanced neuro-inflammatory processes in neurodegenerative diseases and the action of nitrones as potential therapeutics. J Neural Transm Suppl 2000; 60:387-414.
72. Barone FC, Feuerstein GZ. Inflammatory mediators and stroke: New opportunities for novel therapeutics. J Cereb Blood Flow Metab 1999; 19(8):819-834.
73. Rothwell N, Allan S, Toulmond S. The role of interleukin 1 in acute neurodegeneration and stroke: Pathophysiological and therapeutic implications. J Clin Invest 1997; 100(11):2648-2652.
74. Pettmann B, Henderson CE. Neuronal cell death. Neuron 1998; 20(4):633-647.
75. Hartmann A, Mouatt-Prigent A, Faucheux BA et al. FADD: A link between TNF family receptors and caspases in Parkinson's disease. Neurology 2002; 58(2):308-310.
76. Botchkina GI, Geimonen E, Bilof ML et al. Loss of NF-kappaB activity during cerebral ischemia and TNF cytotoxicity. Mol Med 1999; 5(6):372-381.
77. Meistrell IIIrd ME, Botchkina GI, Wang H et al. Tumor necrosis factor is a brain damaging cytokine in cerebral ischemia. Shock 1997; 8(5):341-348.
78. Griffin WS, Sheng JG, Gentleman SM et al. Microglial interleukin-1 alpha expression in human head injury: Correlations with neuronal and neuritic beta-amyloid precursor protein expression. Neurosci Lett 1994; 176(2):133-136.
79. Krupinski J, Kumar P, Kumar S et al. Increased expression of TGF-beta 1 in brain tissue after ischemic stroke in humans. Stroke 1996; 27(5):852-857.
80. Selmaj K, Raine CS, Cannella B et al. Identification of lymphotoxin and tumor necrosis factor in multiple sclerosis lesions. J Clin Invest 1991; 87(3):949-954.
81. Issazadeh S, Ljungdahl A, Hojeberg B et al. Cytokine production in the central nervous system of Lewis rats with experimental autoimmune encephalomyelitis: Dynamics of mRNA expression for interleukin-10, interleukin-12, cytolysin, tumor necrosis factor alpha and tumor necrosis factor beta. J Neuroimmunol 1995; 61(2):205-212.
82. Okuda Y, Nakatsuji Y, Fujimura H et al. Expression of the inducible isoform of nitric oxide synthase in the central nervous system of mice correlates with the severity of actively induced experimental allergic encephalomyelitis. J Neuroimmunol 1995; 62(1):103-112.
83. Renno T, Krakowski M, Piccirillo C et al. TNF-alpha expression by resident microglia and infiltrating leukocytes in the central nervous system of mice with experimental allergic encephalomyelitis. Regulation by Th1 cytokines. J Immunol 1995; 154(2):944-953.

84. Taupin V, Renno T, Bourbonniere L et al. Increased severity of experimental autoimmune encephalomyelitis, chronic macrophage/microglial reactivity, and demyelination in transgenic mice producing tumor necrosis factor-alpha in the central nervous system. Eur J Immunol 1997; 27(4):905-913.
85. Stalder AK, Carson MJ, Pagenstecher A et al. Late-onset chronic inflammatory encephalopathy in immune-competent and severe combined immune-deficient (SCID) mice with astrocyte-targeted expression of tumor necrosis factor. Am J Pathol 1998; 153(3):767-783.
86. Akassoglou K, Bauer J, Kassiotis G et al. Oligodendrocyte apoptosis and primary demyelination induced by local TNF/p55TNF receptor signaling in the central nervous system of transgenic mice: Models for multiple sclerosis with primary oligodendrogliopathy. Am J Pathol 1998; 153(3):801-813.
87. Eugster HP, Frei K, Bachmann R et al. Severity of symptoms and demyelination in MOG-induced EAE depends on TNFR1. Eur J Immunol 1999; 29(2):626-632.
88. Liu J, Marino MW, Wong G et al. TNF is a potent anti- inflammatory cytokine in autoimmune-mediated demyelination. Nat Med 1998; 4(1):78-83.
89. Hjelmstrom P, Juedes AE, Ruddle NH. Cytokines and antibodies in myelin oligodendrocyte glycoprotein-induced experimental allergic encephalomyelitis. Res Immunol 1998; 149(9):794-804, discussion 847-798, 855-760.
90. Wisniewski HM, Wegiel J, Wang KC et al. Ultrastructural studies of the cells forming amyloid in the cortical vessel wall in Alzheimer's disease. Acta Neuropathol 1992; 84:117-127.
91. Selkoe DJ. Altered structural proteins in plaques and tangles: What do they tell us about the biology of Alzheimer's disease? Neurobiol Aging 1986; 7(6):425-432.
92. Dickson DW, Lee SC, Mattiace LA et al. Microglia and cytokines in neurological disease, with special reference to AIDS and Alzheimer's disease. Glia 1993; 7(1):75-83.
93. McGeer PL, Kawamata T, Walker DG et al. Microglia in degenerative neurological disease. Glia 1993; 7(1):84-92.
94. Meda L, Cassatella MA, Szendrei GI et al. Activation of microglial cells by beta-amyloid protein and interferon- gamma. Nature 1995; 374(6523):647-650.
95. Forloni G, Mangiarotti F, Angeretti N et al. Beta-amyloid fragment potentiates IL-6 and TNF-alpha secretion by LPS in astrocytes but not in microglia. Cytokine 1997; 9(10):759-762.
96. Huberman M, Shalit F, Roth-Deri I et al. Correlation of cytokine secretion by mononuclear cells of Alzheimer patients and their disease stage. J Neuroimmunol 1994; 52(2):147-152.
97. Hull M, Fiebich BL, Lieb K et al. Interleukin-6-associated inflammatory processes in Alzheimer's disease: New therapeutic options. Neurobiol Aging 1996; 17(5):795-800.
98. Lieb K, Fiebich BL, Schaller H et al. Interleukin-1 beta and tumor necrosis factor-alpha induce expression of alpha 1-antichymotrypsin in human astrocytoma cells by activation of nuclear factor-kappa B. J Neurochem 1996; 67(5):2039-2044.
99. Fiala M, Zhang L, Gan X et al. Amyloid-beta induces chemokine secretion and monocyte migration across a human blood—brain barrier model. Mol Med 1998; 4(7):480-489.
100. Schubert P, Ogata T, Miyazaki H et al. Pathological immuno-reactions of glial cells in Alzheimer's disease and possible sites of interference. J Neural Transm Suppl 1998; 54:167-174.
101. Hauss-Wegrzyniak B, Dobrzanski P, Stoehr JD et al. Chronic neuroinflammation in rats reproduces components of the neurobiology of Alzheimer's disease. Brain Res 1998; 780(2):294-303.

102. Sutton ET, Thomas T, Bryant MW et al. Amyloid-beta peptide induced inflammatory reaction is mediated by the cytokines tumor necrosis factor and interleukin-1. J Submicrosc Cytol Pathol 1999; 31(3):313-323.
103. Perry RT, Collins JS, Wiener H et al. The role of TNF and its receptors in Alzheimer's disease. Neurobiol Aging 2001; 22(6):873-883.
104. Perini G, Della-Bianca V, Politi V et al. Role of p75 neurotrophin receptor in the neurotoxicity by beta-amyloid peptides and synergistic effect of inflammatory cytokines. J Exp Med 2002; 195(7):907-918.
105. Ebadi M, Bashir RM, Heidrick ML et al. Neurotrophins and their receptors in nerve injury and repair. Neurochem Int 1997; 30(4-5):347-374.
106. Heese K, Hock C, Otten U. Inflammatory signals induce neurotrophin expression in human microglial cells. J Neurochem 1998; 70(2):699-707.
107. Bruunsgaard H, Andersen-Ranberg K, Jeune B et al. A high plasma concentration of TNF-alpha is associated with dementia in centenarians. J Gerontol A Biol Sci Med Sci 1999; 54(7):M357-364.
108. Alvarez XA, Franco A, Fernandez-Novoa L et al. Blood levels of histamine, IL-1 beta, and TNF-alpha in patients with mild to moderate Alzheimer disease. Mol Chem Neuropathol 1996; 29(2-3):237-252.
109. Cacabelos R, Alvarez XA, FrancoMaside A et al. Serum tumor necrosis factor (TNF) in Alzheimer's disease and multi- infarct dementia. Methods Find Exp Clin Pharmacol 1994; 16(1):29-35.
110. Wang J, Asensio VC, Campbell IL. Cytokines and chemokines as mediators of protection and injury in the central nervous system assessed in transgenic mice. Curr Top Microbiol Immunol 2002; 265:23-48.
111. Barger SW, Horster D, Furukawa K et al. Tumor necrosis factors alpha and beta protect neurons against amyloid beta-peptide toxicity: Evidence for involvement of a kappa B-binding factor and attenuation of peroxide and Ca2+ accumulation. Proc Natl Acad Sci USA 1995; 92(20):9328-9332.
112. Weggen S, Eriksen JL, Das P et al. A subset of NSAIDs lower amyloidogenic Abeta42 independently of cyclooxygenase activity. Nature 2001; 414(6860):212-216.
113. Akiyama H, Barger S, Barnum S et al. Inflammation and Alzheimer's disease. Neurobiol Aging 2000; 21(3):383-421.
114. Golde TE. Inflammation takes on Alzheimer disease. Nat Med 2002; 8(9):936-938.
115. Kaires P. NSAID and the ancient defense. Neurology 2002; 59(8):1293, author reply 1293.
116. Xiang Z, Ho L, Yemul S et al. Cyclooxygenase-2 promotes amyloid plaque deposition in a mouse model of Alzheimer's disease neuropathology. Gene Expr 2002; 10(5-6):271-278.
117. Pasinetti GM. From epidemiology to therapeutic trials with anti-inflammatory drugs in Alzheimer's disease: The role of NSAIDs and cyclooxygenase in beta- amyloidosis and clinical dementia. J Alzheimers Dis 2002; 4(5):435-445.
118. Wallach D, Boldin M, Varfolomeev E et al. Cell death induction by receptors of the TNF family: Towards a molecular understanding. FEBS Lett 1997; 410(1):96-106.
119. Lavine SD, Hofman FM, Zlokovic BV. Circulating antibody against tumor necrosis factor-alpha protects rat brain from reperfusion injury. J Cereb Blood Flow Metab 1998; 18(1):52-58.
120. Buttini M, Appel K, Sauter A et al. Expression of tumor necrosis factor alpha after focal cerebral ischaemia in the rat. Neuroscience 1996; 71(1):1-16.
121. Liu T, Clark RK, McDonnell PC et al. Tumor necrosis factor-alpha expression in ischemic neurons. Stroke 1994; 25(7):1481-1488.

122. Hallenbeck JM. Cytokines, macrophages, and leukocytes in brain ischemia. Neurology 1997; 49(5 Suppl 4):S5-9.
123. Nawashiro H, Martin D, Hallenbeck JM. Neuroprotective effects of TNF binding protein in focal cerebral ischemia. Brain Res 1997; 778(2):265-271.
124. Gary DS, Bruce-Keller AJ, Kindy MS et al. Ischemic and excitotoxic brain injury is enhanced in mice lacking the p55 tumor necrosis factor receptor. J Cereb Blood Flow Metab 1998; 18(12):1283-1287.
125. Wong GH. Protective roles of cytokines against radiation: Induction of mitochondrial MnSOD. Biochim Biophys Acta 1995; 1271(1):205-209.
126. Fineman I, Hovda DA, Smith M et al. Concussive brain injury is associated with a prolonged accumulation of calcium: A 45Ca autoradiographic study. Brain Res 1993; 624(1-2):94-102.
127. Nilsson P, Laursen H, Hillered L et al. Calcium movements in traumatic brain injury: The role of glutamate receptor-operated ion channels. J Cereb Blood Flow Metab 1996; 16(2):262-270.
128. Appel E, Kolman O, Kazimirsky G et al. Regulation of GDNF expression in cultured astrocytes by inflammatory stimuli. Neuroreport 1997; 8(15):3309-3312.
129. Aloisi F, Borsellino G, Care A et al. Cytokine regulation of astrocyte function: In-vitro studies using cells from the human brain. Int J Dev Neurosci 1995; 13(3-4):265-274.
130. Hattori A, Tanaka E, Murase K et al. Tumor necrosis factor stimulates the synthesis and secretion of biologically active nerve growth factor in nonneuronal cells. J Biol Chem 1993; 268(4):2577-2582.
131. Lang AE, Lozano AM. Parkinson's disease. Second of two parts. N Engl J Med 1998; 339(16):1130-1143.
132. Lang AE, Lozano AM. Parkinson's disease. First of two parts. N Engl J Med 1998; 339(15):1044-1053.
133. Wintermeyer P, Riess O, Schols L et al. Mutation analysis and association studies of nuclear factor-kappaB1 in sporadic Parkinson's disease patients. J Neural Transm 2002; 109(9):1181-1188.
134. Rousselet E, Callebert J, Parain K et al. Role of TNF-alpha receptors in mice intoxicated with the parkinsonian toxin MPTP. Exp Neurol 2002; 177(1):183-192.
135. McGuire SO, Ling ZD, Lipton JW et al. Tumor necrosis factor alpha is toxic to embryonic mesencephalic dopamine neurons. Exp Neurol 2001; 169(2):219-230.
136. Mogi M, Togari A, Kondo T et al. Glial cell line-derived neurotrophic factor in the substantia nigra from control and parkinsonian brains. Neurosci Lett 2001; 300(3):179-181.
137. Sriram K, Matheson JM, Benkovic SA et al. Mice deficient in TNF receptors are protected against dopaminergic neurotoxicity: Implications for Parkinson's disease. Faseb J 2002; 16(11):1474-1476.
138. Nagatsu T, Mogi M, Ichinose H et al. Changes in cytokines and neurotrophins in Parkinson's disease. J Neural Transm Suppl 2000; 60:277-290.
139. Ruberg M, France-Lanord V, Brugg B et al. [Neuronal death caused by apoptosis in Parkinson disease]. Rev Neurol (Paris) 1997; 153(8-9):499-508.
140. Kaltschmidt B, Baeuerle PA, Kaltschmidt C. Potential involvement of the transcription factor NF-kappa B in neurological disorders. Mol Aspects Med 1993; 14(3):171-190.
141. Hartmann A, Troadec JD, Hunot S et al. Caspase-8 is an effector in apoptotic death of dopaminergic neurons in Parkinson's disease, but pathway inhibition results in neuronal necrosis. J Neurosci 2001; 21(7):2247-2255.
142. Viswanath V, Wu Y, Boonplueang R et al. Caspase-9 activation results in downstream caspase-8 activation and bid cleavage in 1-methyl-4-phenyl-1,2,3,6-tetrahydropyridine-induced Parkinson's disease. J Neurosci 2001; 21(24):9519-9528.

143. Mogi M, Togari A, Kondo T et al. Caspase activities and tumor necrosis factor receptor R1 (p55) level are elevated in the substantia nigra from parkinsonian brain. J Neural Transm 2000; 107(3):335-341.
144. Deng HX, Hentati A, Tainer JA et al. Amyotrophic lateral sclerosis and structural defects in Cu,Zn superoxide dismutase. Science 1993; 261(5124):1047-1051.
145. Reaume AG, Elliott JL, Hoffman EK et al. Motor neurons in Cu/Zn superoxide dismutase-deficient mice develop normally but exhibit enhanced cell death after axonal injury. Nat Genet 1996; 13(1):43-47.
146. Wong PC, Pardo CA, Borchelt DR et al. An adverse property of a familial ALS-linked SOD1 mutation causes motor neuron disease characterized by vacuolar degeneration of mitochondria. Neuron 1995; 14(6):1105-1116.
147. Estevez AG, Crow JP, Sampson JB et al. Induction of nitric oxide-dependent apoptosis in motor neurons by zinc-deficient superoxide dismutase. Science 1999; 286(5449):2498-2500.
148. Lyons TJ, Liu H, Goto JJ et al. Mutations in copper-zinc superoxide dismutase that cause amyotrophic lateral sclerosis alter the zinc binding site and the redox behavior of the protein. Proc Natl Acad Sci USA 1996; 93(22):12240-12244.
149. Wiedau-Pazos M, Goto JJ, Rabizadeh S et al. Altered reactivity of superoxide dismutase in familial amyotrophic lateral sclerosis [see comments]. Science 1996; 271(5248):515-518.
150. Bruijn LI, Beal MF, Becher MW et al. Elevated free nitrotyrosine levels, but not protein-bound nitrotyrosine or hydroxyl radicals, throughout amyotrophic lateral sclerosis (ALS)- like disease implicate tyrosine nitration as an aberrant in vivo property of one familial ALS-linked superoxide dismutase 1 mutant. Proc Natl Acad Sci USA 1997; 94(14):7606-7611.
151. Ripps ME, Huntley GW, Hof PR et al. Transgenic mice expressing an altered murine superoxide dismutase gene provide an animal model of amyotrophic lateral sclerosis. Proc Natl Acad Sci USA 1995; 92(3):689-693.
152. Wong PC, Waggoner D, Subramaniam JR et al. Copper chaperone for superoxide dismutase is essential to activate mammalian Cu/Zn superoxide dismutase. Proc Natl Acad Sci USA 2000; 97(6):2886-2891.
153. Subramaniam JR, Lyons WE, Liu J et al. Mutant SOD1 causes motor neuron disease independent of copper chaperone- mediated copper loading. Nat Neurosci 2002; 5(4):301-307.
154. Cleveland DW, Rothstein JD. From Charcot to Lou Gehrig: Deciphering selective motor neuron death in ALS. Nat Rev Neurosci 2001; 2(11):806-819.
155. Taylor JP, Hardy J, Fischbeck KH. Toxic proteins in neurodegenerative disease. Science 2002; 296(5575):1991-1995.
156. Julien JP. Amyotrophic lateral sclerosis. Unfolding the toxicity of the misfolded. Cell 2001; 104(4):581-591.
157. Kopito RR. Aggresomes, inclusion bodies and protein aggregation. Trends Cell Biol 2000; 10(12):524-530.
158. Kawamata T, Akiyama H, Yamada T et al. Immunologic reactions in amyotrophic lateral sclerosis brain and spinal cord tissue. Am J Pathol 1992; 140(3):691-707.
159. McGeer PL, McGeer EG. Inflammatory processes in amyotrophic lateral sclerosis. Muscle Nerve 2002; 26(4):459-470.
160. Ghezzi P, Mennini T. Tumor necrosis factor and motoneuronal degeneration: An open problem. Neuroimmunomodulation 2001; 9(4):178-182.

161. Poloni M, Facchetti D, Mai R et al. Circulating levels of tumour necrosis factor-alpha and its soluble receptors are increased in the blood of patients with amyotrophic lateral sclerosis. Neurosci Lett 2000; 287(3):211-214.
162. Robertson J, Beaulieu JM, Doroudchi MM et al. Apoptotic death of neurons exhibiting peripherin aggregates is mediated by the proinflammatory cytokine tumor necrosis factor-alpha. J Cell Biol 2001; 155(2):217-226.
163. Beaulieu JM, Nguyen MD, Julien JP. Late onset death of motor neurons in mice overexpressing wild-type peripherin. J Cell Biol 1999; 147(3):531-544.
164. Elliott JL. Cytokine upregulation in a murine model of familial amyotrophic lateral sclerosis. Brain Res Mol Brain Res 2001; 95(1-2):172-178.
165. Yoshihara T, Ishigaki S, Yamamoto M et al. Differential expression of inflammation- and apoptosis-related genes in spinal cords of a mutant SOD1 transgenic mouse model of familial amyotrophic lateral sclerosis. J Neurochem 2002; 80(1):158-167.
166. Hensley K, Floyd RA, Gordon B et al. Temporal patterns of cytokine and apoptosis-related gene expression in spinal cords of the G93A-SOD1 mouse model of amyotrophic lateral sclerosis. J Neurochem 2002; 82(2):365-374.

Chapter 5

The Role of LIGHT in Autoimmunity

Jing Wang and Yang-Xin Fu*

Abstract

This chapter focuses on the role of LIGHT in the induction of autoimmunity. LIGHT and LTαβ share the same receptor, LTβR, and cooperate in lymphoid organogenesis and development of lymphoid structure. Previous findings establish a crucial biological role for LIGHT, a T cell-derived costimulatory ligand, in T cell activation and expansion via a T-T cell dependent manner and the dysregulation of LIGHT activity results in the disturbance of T cell homeostasis and ultimately in the breakdown of peripheral tolerance. Furthermore, the blockade of LIGHT activity ameliorates the severity of T cell-mediated diseases indicating the essential involvement of LIGHT in various pathological conditions. Here, we will review the recent studies about LIGHT mainly in the context of autoimmunity and conclude with a discussion of the potential mechanisms by which LIGHT promotes autoimmunity.

Receptor and Ligand Interaction

Members of the TNF/TNFR superfamily play multiple roles in the cellular differentiation, survival, and death pathways that orchestrate lymphoid organogenesis, activation and homeostasis of immune cells.[1] TNF and LTα along with LIGHT and LTβ, define a core group of ligands that bind four cognate cell surface receptors TNFRI, TNFRII, LTβR and HVEM with significant complexities of receptor cross-utilization (Fig. 1). Membrane-bound form of lymphotoxin (LTαβ) and its receptor LTβR have been studied extensively and their essential roles in the development and organization of secondary lymphoid tissues and ectopic lymphoid neogenesis were well established.[2-8] LIGHT, a newly discovered TNF superfamily member (TNFSF14), is a type II transmembrane protein expressed on activated T cells and immature dendritic cells.[9,10] The primary structure of human LIGHT protein predicted from the cDNA sequence contains 240 amino acids. Human LIGHT exhibits significant sequence homology with the

*Corresponding Author: Yang-Xin Fu—The University of Chicago, Department of Pathology and Committee on Immunology, 5841 S. Maryland, Chicago, Illinois 60637, USA. Email: yfu@midway.uchicago.edu

TNF Superfamily, edited by Sanjay Khare. ©2007 Landes Bioscience

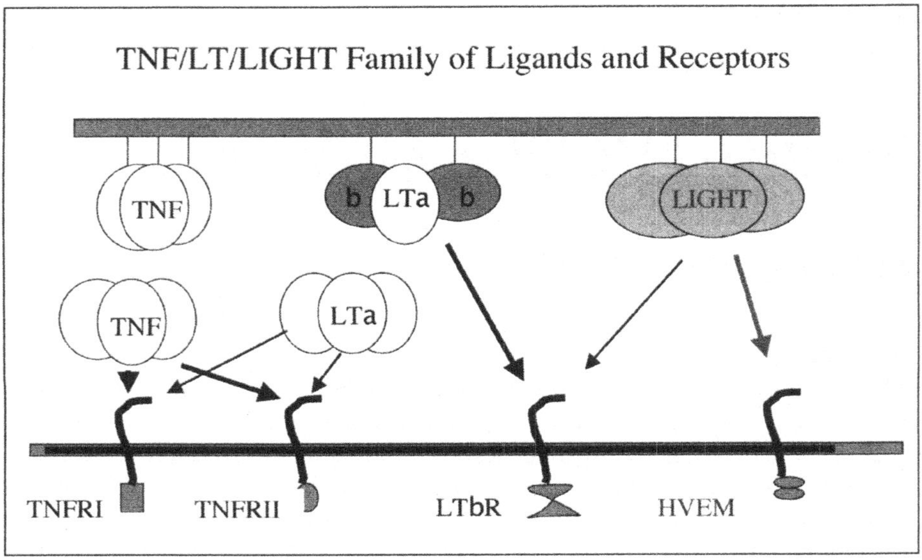

Figure 1. A current model for the LT/LIGHT family. LTβR binds to both membrane LTαβ and LIGHT while HVEM binds to LIGHT and soluble $LT\alpha_3$. Therefore, LIGHT binds to both LTβR and HVEM. Soluble $TNF\alpha_3$ and $LT\alpha_3$ bind to TNFRI and TNFRII.

C-terminal receptor-binding domains of LTβ (34% identity), Fas Ligand (31%), 4-1BBL (29%), TRAIL (28%), LTα (27%), TNF (27%), and CD40L (26%).[9] From the mouse cDNA for LIGHT, a protein of 239 amino acids can be deduced, with characteristics of a type II transmembrane protein and 77% amino acid homology with human LIGHT.[11] The expected receptor-binding region of mouse LIGHT has substantial sequence homology with those of Fas ligand (33%), LTβ (30%), LTα (28%), TNF (27%), receptor activator of nuclear factor-κB ligand (26%) and TNF-related apoptosis-inducing ligand (23%).[11] LIGHT can bind three receptors;[9] LTβR expressed on stromal cells and nonlymphoid hematopoietic cells,[4,12] HVEM expressed on T, B and other hematopoietic cells,[13-15] and DcR3, a decoy receptor which also binds to Fas ligand (FasL).[16]

Many members of the ligands in the TNF superfamily appear to be involved in regulating T cell homeostasis, which is reflected in the highly conserved genomic organization of these ligands (Fig. 2). The *LIGHT* gene is mapped to human chromosome 19 and clustered with 4-1BB ligand and CD27 ligand.[17] The LIGHT locus shares striking similarity in organization to the TNF superfamily locus, in which TNF, LTα and LTβ are closely clustered residing within the major histocompatibility complex (MHC) on human chromosome 6.[18] It is known that the TNF superfamily members in these paralogous gene clusters, such as the ligands for CD27, 4-1BB, and OX40,[19,20] function as costimulatory molecules enhancing T lymphocyte activation and survival, or induce elimination of activated T

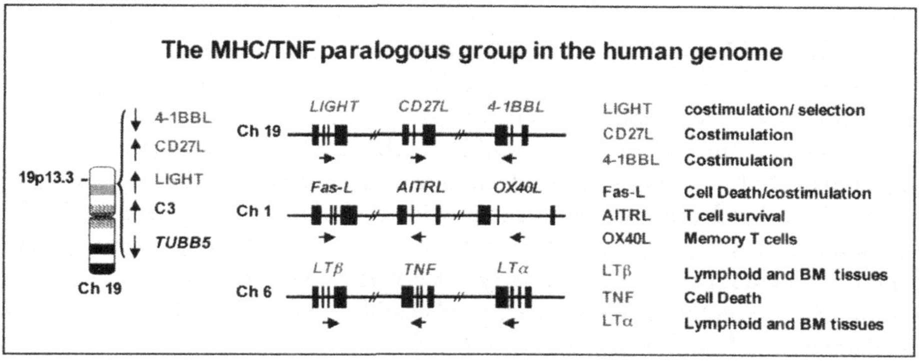

Figure 2. Organization of the TNF superfamily genes within the paralogous regions present on chromosomes 1, 6, and 19. Left panel: Diagram of the human chromosome 19p13.3 region containing the LIGHT genomic locus. Middle panel: Distribution of TNF paralogous superfamily gene clusters arranged from centromere to telomere. Arrows indicate gene transcriptional orientation, and solid blocks represent exons. LIGHT is 7.78 kb from *C3* and about 79 kb from *CD27L*. *CD27L* is about 235 kb from *4-1BBL*. *FasL* is separated from *AITRL* by 374 kb, while *AITRL* and *OX40L* are 134 kb apart. *TNF* is 2.9 kb from *LT*b and 1.3 kb from *LT*a. Right panel: activities of the TNF-related ligands on T cells. Modified and reproduced with permission from Dr. Carl Ware. Also see Granger SW and Ware CF. Turning on LIGHT. J Clin Invest 2001; 108:1741-1742.

cells, as described for TNF and FasL.[1] It is likely that the evolutionary conservation of the TNF related ligands dedicated to T cell homeostasis and linkage to antigen recognition molecules reflects their importance in fine-tuning antigen recognition and immune tolerance.

The Role of LIGHT in T Cell Activation

T Cell-Derived LIGHT Functions as a Costimulatory Molecule for Expansion of T Cells

Costimulatory molecules on antigen presenting cells (APCs) play an important role in T cell activation and expansion. The well characterized costimulatory pathway for optimal T cell activation involves the T cell surface molecule CD28, which responds to the costimulatory molecules B7-1 (CD80) and B7-2 (CD86) expressed on activated antigen-presenting cells (APCs).[21] Previous studies have shown that murine B7 molecules could costimulate with anti-CD3 monoclonal antibody (mAb) or concanavalin A (ConA) to induce T cell activation.[22-24] Anti-CD3 mAb can directly crosslink the TCR complex and stimulate T cell proliferation in an APC independent way whereas ConA induces T cell activation via an APC-dependent mechanism.[25,26] $CD28^{-/-}$ mice have impaired responsiveness to ConA suggesting that the interaction of B7 and CD28 is critical for the

APC-dependent T cell activation.[27] CTLA4-Ig, a soluble receptor for B7, could block ConA and anti-CD3 mAb induced proliferation in splenocytes or lymph node cells.[27-29] However, cultures of T cells that had been rigorously depleted of accessory cells were found to proliferate in a B7-independent manner.[29] These experiments, therefore, raise two possibilities that an APC-derived costimulatory signal may not be necessary under all circumstances such as direct cross-linking of the TCR, or that T cells may be able to provide costimulation to each other via the ligand(s) and receptor(s) expressed on T cells themselves. However, it is unclear whether such additional costimulatory molecules are present and whether the ligation of these molecules by T cell-derived costimulatory ligand(s) is required for further activation and/or expansion of T cells.

Our recent studies demonstrated that the blockade of LIGHT by its soluble receptor HVEM-Ig dramatically reduced the anti-CD3 mediated T cell proliferation in the absence of APCs indicating that LIGHT can function as a costimulatory molecule for the complete expansion of peripheral T cells in a T-T cell dependent manner.[30] In contrast to reagents that block LIGHT activity, CTLA4-Ig did not show any impact on the proliferation of T cells in our APC-free system.[30] These results are consistent with the notion that CD28 interactions with the B7 family of costimulatory ligands are essential for inducing T cell activation via an APC-dependent mechanism[21,31,32] while LIGHT might be important for T-T cell interaction. Taken together, these results support our hypothesis that LIGHT from T cells is required for T cell expansion via T-T cell interaction whereas B7-1/B7-2 from APCs are probably more important for initiating T cell responses during the early priming phase.

Other studies have also shown that LIGHT has potent, CD28-independent costimulatory activity and results in enhanced T cell proliferation and secretion of gamma interferon (IFN-γ) and granulocyte-macrophage colony-stimulating factor (GM-CSF) in vitro.[10,11] Although we emphasize the role of T cell-derived LIGHT in activation of T cells, Tamada et al[10] reported that blockade of LIGHT by its soluble receptors, LTβR-Ig or HVEM-Ig, inhibits the induction of DC-mediated primary allogeneic T cell response suggesting that LIGHT may function as a costimulatory molecule in DC-mediated cellular immune responses. However, whether the costimulatory activity of LIGHT is derived from T cells or DC is unclear in that model. Furthermore, engagement of LIGHT amplifies the NF-kappaB signaling pathway, and preferentially induces the production of IFN-gamma, but not IL-4, in the presence of an antigenic signal.[10]

Impaired T Cell Activation in LIGHT Knockout Mice Indicated an Essential Role of LIGHT in T Cell Response

Gene targeting approaches have largely confirmed in vitro data regarding the costimulatory activity of LIGHT. LIGHT$^{-/-}$ mice showed a reduced cytotoxic T lymphocyte (CTL) activity and cytokine production in allogeneic mixed lymphocytes reaction (MLR) studies.[33] Detailed analysis revealed that proliferative responses of CD8$^+$ T cells are impaired and interleukin 2 (IL-2) production of CD4$^+$

T cells is defective in the absence of LIGHT.[33] Furthermore, a reduced ^{3}[H]-thymidine incorporation after TCR stimulation was observed for LIGHT$^{-/-}$ T cells.[33] Collectively, these results indicate LIGHT has important costimulatory functions for T cell activation. An independent study also showed that Vβ8^{+}CD8^{+} T cell proliferation in response to staphylococcal enterotoxin B (SEB) was significantly reduced in LIGHT$^{-/-}$ mice. Consistently, induction and cytokine secretion of CD8^{+} CTL to MHC class I-restricted peptide was impaired in LIGHT$^{-/-}$ mice. However, the proliferative response of Vβ8^{+}CD4^{+} T cells to SEB was comparable in LIGHT$^{-/-}$ and LIGHT$^{+/+}$ mice in this report. Thus, they proposed that LIGHT is required for proliferation of normal CD8^{+} T cells but not CD4^{+} T cells.[34]

The Role of LIGHT in Systemic Autoimmunity

Results from in vitro culture models appear to support a role for LIGHT in T cell activation,[11,13] the phenotype of mice overexpressing LIGHT provides evidence that up-regulation of LIGHT can play a critical role in T cell mediated inflammation and autoimmune diseases. The studies from our group[30] and Shaikh et al[35] demonstrate that constitutive expression of LIGHT results in multiorgan inflammation caused by activated T cells. Normally, LIGHT is transiently expressed on the surface of T cells following activation[9] and downregulated upon the termination of immune responses, but in the studies discussed here, two different lineage-specific promoters were used to drive the constitutive expression of LIGHT in T cells which eventually leads to breakdown of peripheral tolerance and development of autoimmune syndromes.

To investigate the role of T cell-derived LIGHT in the expansion of T cells in vivo, our group generated a transgenic line that constitutively expresses the LIGHT protein under the control of the proximal *lck* promoter and CD2 enhancer, which gives rise to a T cell lineage-specific expression of LIGHT.[36,37] Lck-LIGHT Tg mice spontaneously develop severe autoimmune disease manifested by splenomegaly, lymphadenopathy, glomerulonephritis, elevated autoantibodies, and severe infiltration of various peripheral tissues.[30] In contrast to mice transgenic for BAFF, another TNF family member, which had enlarged secondary lymphoid tissues due to the expanded B cell compartment,[38-40] most expansion occurred in the T cell compartment of LIGHT Tg mice.[30,35] These data strongly support the hypothesis that T cell-derived LIGHT is sufficient to cause the expansion of peripheral T cells in vivo.

Apart from the significantly enlarged and hyperactivated T cell compartment, augmented cytokine production and expansion of granulocyte-macrophage lineage in the spleen were also observed in lck-LIGHT Tg mice.[30] IFN-γ producing T cells were significantly increased in Tg mice further demonstrating the sufficiency of T cell-derived LIGHT to induce hyperactivation of T cells in vivo.[30] GM-CSF promotes hematopoiesis and leads to the enlargement of the spleen with the preferential increase of GM lineages. Although higher GM-CSF production was detected in LIGHT Tg mice, the source of increased GM-CSF remains to be determined. We speculate that constitutive expression of T cell-derived

LIGHT may enhance the production of GM-CSF from activated T cells leading to systemic hematopoiesis in the Tg mice. Further analysis indeed showed an expansion of macrophage and granulocyte populations in the spleen of Tg mice.[30] Macrophage is one of the major cellular components involved in chronic inflammation and autoimmunity largely owing to its proinflammatory cytokine network,[41,42] thus it would be of great interest to dissect the mechanism by which LIGHT elicits the activation and expansion of macrophage. We predict it could be mediated either by activated T cell-derived IFN-γ or direct ligand/receptor interaction on T cells and macrophages.

Lck-LIGHT Tg mice developed severe autoimmune manifestations.[30] Striking phenotypes were consistently observed in lck-LIGHT Tg mice beginning at 5 months, suggesting the crucial role of LIGHT in the induction of autoimmunity. Microscopic examination of lck-LIGHT Tg mice revealed the dramatic inflammatory cell infiltrate in the lamina propria and submucosa of intestine with prominent germinal center formation in a diffuse pattern.[30] In addition, severe cutaneous lesions along with ulceration and scar formation were observed in the aged transgenic mice. Histological sections demonstrated conspicuous mixed acute and chronic inflammatory cell infiltrate extending from epidermis to subcutis in lck-LIGHT Tg mice.[30] More intriguing phenotypes were revealed by renal pathological analysis in lck-LIGHT Tg mice which spontaneously developed diffuse global proliferative glomerulonephritis involving over 80% of the glomeruli. Consistent with this observation, immunofluorescence staining revealed strong diffuse IgG deposition in a coarsely granular pattern in Tg mice, similar to what is often observed in type IV lupus patients. Immunofluorescence staining against total immunoglobulin light chains showed intense positive staining in a similar pattern as IgG.[30] Elevation of autoantibodies serves as criteria for the clinical diagnosis of autoimmune disease and has been shown to be characteristic for MRL-*lpr/lpr* mice.[43] LIGHT Tg mice demonstrated elevated anti-DNA autoantibodies and rheumatoid factors (RF), another commonly detected autoantibody in chronic inflammation and autoimmune diseases. It appears that the phenotypes observed in lck-LIGHT Tg mice share certain similarity with those in MRL-*lpr/lpr* mice, an established murine model for systemic lupus erythematosus (SLE), which may be attributed to the critical roles of both TNF-related ligands in the regulation of T cell homeostasis and its disturbance leads to lymphoproliferative disorder and autoimmune diseases.

The findings of lupus-like glomerulonephritis, increased inflammatory cell infiltrate in multiple organs, along with elevations of serum autoantibodies indicated the establishment of autoimmunity in lck-LIGHT Tg mice.[30] Therefore, the overproliferation and hyperactivation of T cells mediated by T cell-derived LIGHT resulted in the breakdown of T/B cell tolerance, supporting the notion that the dysregulation of LIGHT expression may be a critical element in the induction of both T and B cell autoimmunity and in the pathogenesis of autoimmune diseases.

Studies from CD2-LIGHT Tg mice in which constitutive LIGHT expression was driven by CD2 promoter and enhancer showed lymphoid tissue abnormalities, including splenomegaly, lymphadenopathy, and pronounced inflammation in the intestine, consisting of expanded populations of conventional $CD4^+$ and $CD8^+\alpha\beta$ T cells.[35] The inflamed intestines displayed signs of chronic inflammation including loss of goblet cells, distortion and hyperplasia of crypts, villous atrophy, and mononuclear cell infiltrates. Thus, increased or sustained expression of LIGHT on activated T cells contributes to the induction and persistence of inflammation in the intestine demonstrated by both of the transgenic lines.[30,35]

However, the two studies showed some differences in the phenotypes of the LIGHT transgenic mice.[30,35] Mice expressing LIGHT under the *lck* promoter exhibit severe inflammation in the skin and moderate inflammation in kidney in addition to the intestine, whereas mice with *CD2*-LIGHT fail to reproduce because of severely atrophied reproductive organs. The autoimmune nephritis, accompanied with anti-DNA antibodies, seen in *lck*-LIGHT mice suggests that LIGHT induces a loss of self-tolerance.[30] These phenotypic differences most likely reflect differences in tissue-specific expression by these promoters, since CD2 is expressed in T and some B cells, whereas lck is active in thymocytes and peripheral T cells. Quantitative differences in ligand expression, and effects on different populations of cells may also help explain discrepancies between the two reports.[44] Nevertheless, both studies provide convincing evidence that LIGHT plays a key role in T cell homeostasis and peripheral tolerance.

Results from recent studies demonstrate that LIGHT is an important costimulatory molecule functioning in a T-T cell dependent manner required for the complete expansion of peripheral T cells. The dysregulation or overexpression of LIGHT may play a key role in the pathogenesis of T cell-mediated inflammation and autoimmunity. Furthermore, transgenic model indicates that LIGHT is sufficient to cause the activation and expansion of peripheral T cells that subsequently lead to the breakdown of peripheral tolerance. LIGHT transgenic model brings new insight into the pathogenesis of various autoimmune disorders and provides an interesting framework for studying the mechanisms regulating T cell activation, immune tolerance and the induction of autoimmunity. In the following sections, we will discuss that upregulation of LIGHT may contribute to the pathogenesis of T cell-mediated diseases.

The Role of LIGHT in T Cell-Mediated Disease Model

Emerging evidence indicates that LIGHT is a key player in T cell homeostasis and peripheral tolerance. Studies by Wang et al[30] and Shaikh et al[35] reveal that sustained expression of LIGHT can cause profound inflammation and loss of tolerance leading to autoimmune syndromes. These new findings validate LIGHT as an important T cell regulatory molecule and suggest its candidacy as a pharmaceutical target for diseases involving T cells.

Type I Diabetes

Insulin-dependent diabetes mellitus (IDDM) is a T cell-mediated autoimmune disease, in which the insulin-producing beta cells are selectively destroyed by autoreactive T cells and the nondiabetic (NOD) mouse is the well-established model for studies of IDDM.[45,46] Previous studies suggested that the administration of LTβR-Ig (a chimera of the receptor's ligand-binding domain fused with the Fc region of IgG that neutralizes both LIGHT and LTαβ) blocked the development of IDDM[47] and an independent study from LTβR-Fc Tg mice also supported the role of LTβR in IDDM.[48] Since membrane LTαβ and LIGHT both bind to LTβR, the therapeutic effects of LTβR-Ig treatment could be attributed to either or both ligands. One striking feature of spontaneous autoimmune diabetes is the prototypic formation of lymphoid follicular structures within the pancreas and membrane LTαβ has been shown to play an important role in the formation of lymphoid tissues, therefore, it was proposed that membrane LTαβ involved in the development of type I diabetes.[47] The mechanisms by which membrane LTαβ contributes to type I diabetes largely reside in its ability of promoting the formation of lymphoid microenvironment required for the development and progression of IDDM.[47] It is possible that LIGHT can contribute to the development of lymphoid tissue for IDDM since upregulation of LIGHT can stimulate LTβR and induce the formation of lymphoid structures in the absence of LT.[49]

To study whether LIGHT is involved in the development of autoimmune diabetes, HVEM-Ig, a soluble receptor for LIGHT, was used to neutralize LIGHT signaling in NOD mice. At the age of 6-7 weeks, many islets in NOD mice were already infiltrated with autoreactive T cells and treatment with HVEM-Ig at this time significantly prevented the development of IDDM and reduced the incidence of diabetes (80% in control vs. 25% in treated group).[30] HVEM is a receptor for LIGHT and does not bind to membrane LTαβ although shows very weak binding to $LT\alpha_3$.[9] These results suggest that the blockade of LIGHT by HVEM-Ig prevents the pathogenesis of IDDM and LIGHT may play a critical role in the development of type I diabetes.[30] However, there are several unresolved issues. Earlier studies showed that LTβR-Ig treatment prevented the development of IDDM induced by diabetogenic T cells in an adoptive transfer model and similar approach should be applied to HVEM-Ig treatment to test the role of LIGHT in different phases of IDDM progression. The development of insulitis needs to be addressed in HVEM-Ig treatment to determine if LIGHT is an effector molecule in the tissue destructive phase of IDDM. Moreover, the effect of anti-LTβ antibody, which only blocks the membrane LTαβ signaling, should be examined to distinguish the impact of two ligands for LTβR in the IDDM. We predict that administration of LTβR-Ig, which blocks both ligands, probably has more potent therapeutic effect on the type I diabetes than the blockade of either ligand.

Transplantation and Tumor Rejection

The effect of LIGHT in transplantation was first examined in a graft-versus-host disease (GVHD) model.[11] Blockade of LIGHT by administration of soluble receptor

LTβR-Ig or neutralizing antibody against LIGHT led to ameliorated GVHD. When $LT\alpha^{-/-}$ mice were used as recipients lacking both soluble $LT\alpha_3$ and membrane LTαβ, the therapeutic effect of LTβR-Ig persisted in this GVHD model, which strongly argued the critical role of LIGHT in the development of GVHD.[11] Chen's group has demonstrated that infusion of an mAb against CD40 ligand (CD40L) further increases the efficacy of LTβR-Ig, leading to complete prevention of GVHD and tolerance.[50]

The role of LIGHT-HVEM costimulation was examined in a murine cardiac allograft rejection model.[51] Allografts upregulated the expression of LIGHT and HVEM on infiltrating leukocytes starting from 3 days after transplantation although normal hearts lacked both LIGHT and HVEM mRNA expression. There was no significant difference between the mean survival of fully MHC-mismatched cardiac allografts in $LIGHT^{-/-}$ mice, cyclosporine A (CsA)-treated $LIGHT^{+/+}$ or $LIGHT^{+/+}$ mice. In contrast, mean survival of allograft in CsA-treated $LIGHT^{-/-}$ recipients was considerably prolonged compared with either untreated $LIGHT^{-/-}$ or CsA-treated $LIGHT^{+/+}$ mice. The beneficial effects of the deletion of LIGHT in CsA-treated recipients were associated with the reduction of IFN-γ, inducible protein-10 (IFN-γ-induced chemokine), and its receptor CXCR3 in the allografts.[51] Consistently, it has been reported earlier that DcR3/TR6, a soluble decoy receptor for LIGHT, can also delay the onset of cardiac allograft rejection.[52] These data suggest that T cell to T cell-mediated LIGHT/HVEM-dependent costimulation is a significant component of the host response mediating cardiac allograft rejection. In addition to its impact in cardiac rejection model, LIGHT has been shown to act synergistically with CD28 in skin allograft rejection in vivo.[33]

Gene transfer of LIGHT into tumor nodules induced an antigen-specific cytotoxic T lymphocytes (CTLs) response to tumor antigens and therapeutic immunity against established mouse P815 tumor.[11] Depletion of $CD8^+$ T cells completely abrogated the anti-tumor effect of LIGHT, whereas the anti-tumor effect was partially inhibited by depletion of the $CD4^+$ T cell.[11] These results indicate that LIGHT costimulation in vivo can enhance the CTL response to tumor antigen and eradicate tumors via a T cell-dependent mechanism.

Inflammatory Bowel Disease (IBD)

Human inflammatory bowel disease (IBD) is a chronic, relapsing and remitting inflammatory condition of unknown origin. Clinical and basic studies in humans with IBD have long suggested that genetic and environmental factors play an inter-related role in the pathogenesis of this disorder. Recent immunologic studies of the disease indicated that IBD is due to a dysregulated mucosal immune response to one or more unknown antigens present in the normal, indigenous bacterial flora.[53] Various animal models of IBD have been helpful in the dissection of the mechanisms involved in IBD pathogenesis.

An experimental model of mucosal inflammation has been produced by creating mice that overexpress TNF (TNFΔARE mice).[54] The intestinal pathology in this model results from TNF signaling through either TNFRI (p55) or TNFRII (p75). In addition to TNF, other members of the TNF family can contribute to the development of experimental mucosal inflammation. For example, the importance of CD40-CD40L in mucosal inflammation is shown by the fact that administration of anti-CD40-ligand antibody completely blocks trinitrobenzene sulphonic acid (TNBS)-induced colitis.[55] Similarly, LTδR-Ig prevents colitis in a CD4 T cell-dependent transfer model.[6] Furthermore, overexpression of LIGHT leads to development of intestinal inflammation,[30,35] which implicates the critical role of LIGHT in IBD pathogenesis.

Interestingly, lck-LIGHT Tg mice appear to have active mononuclear cellular infiltration in many organs including the intestine and the skin which show the most evident pathological manifestations such as ulceration and scar formation in the skin and massive infiltrate in the intestine.[30] It is possible that the generalized expansion of activated T cells is harmless to the host except in tissues that interface with the external environment, where excessive T cell response to microbial or environmental antigens results in local pathology. In other words, the autoimmune disease may represent dysregulated homeostasis in response to environmental antigens.

Overall, these studies indicate that LIGHT is involved in T cell-mediated diseases and its dysregulation may trigger the abnormal activation of T cells, spawning severe tissue destruction and autoimmune manifestations. Thus, beneficial effects may be obtained by blockade of LIGHT upregulation in autoimmune diseases and GVHD. In contrast, the enhancement of LIGHT expression may be desired in tumor rejection. Animal models clearly demonstrate that dysregulated expression of the TNF-related cytokines leads to severe immunopathology. These new findings help clear the path to therapeutic interventions of autoimmune diseases and tumor rejection in the future.

The Potential Mechanism for LIGHT-Mediated Autoimmunity

Is the tissue destruction observed in the LIGHT transgenic mice due to nonspecific inflammation by activated T cells or true autoimmunity due to loss of tolerance? The current data seem to suggest that there is a loss of self-tolerance but the detailed mechanism remains to be determined. Interestingly, in both types of LIGHT transgenic mice, the size of the thymi is remarkably reduced and less CD4/CD8 double-positive (DP) cells are observed. Since DP thymocytes are normally subject to negative selection, these results raise the possibility that LIGHT might be involved in negative selection.[35,56] Moreover, our study showed that blockade of LIGHT signaling in vitro and in vivo prevented negative selection induced by intrathymically expressed antigens, resulting in the rescue of thymocytes from apoptosis.[56] Although speculation abounds, no other TNF family member has yet been confirmed as a factor in modulating negative selection.[57] The current studies[35,56] suggest that LIGHT affects central differentiation processes critical for T cell tolerance. However, LIGHT deficient mice showed no obvious defects in thymus,[33,34] thus experiments designed to test whether central tolerance is affected by the absence of LIGHT or whether the LIGHT-mediated thymocyte

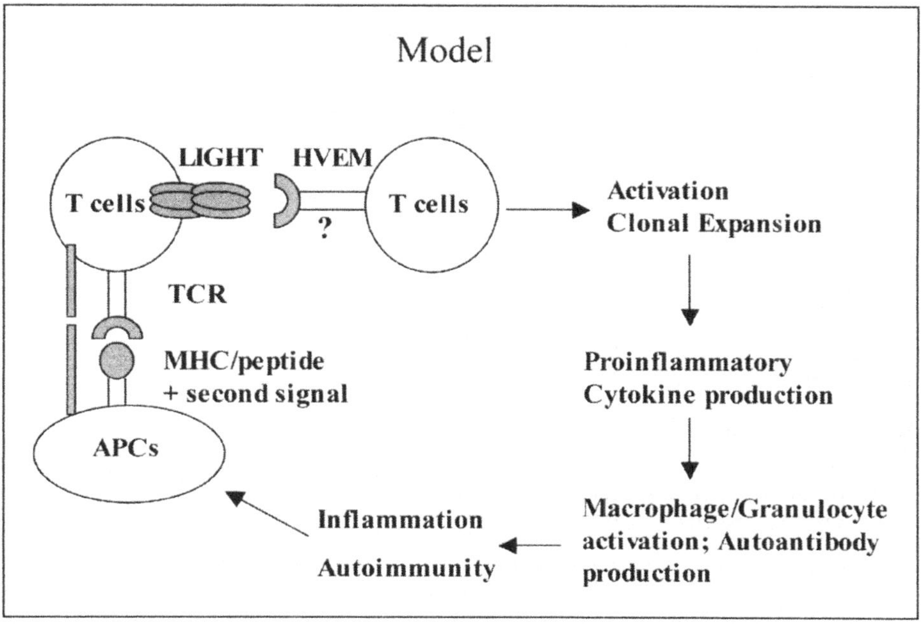

Figure 3. Proposed model for the LIGHT-induced autoimmunity. Question mark means other unidentified receptor(s).

deletion is dependent on the interaction between TCR and self-MHC/peptide probably will provide more insights into this issue.

The existence of central tolerance implies that immature thymocytes respond differently to the antigen encountered than do mature T cells. Our findings with LIGHT provide an example of a T cell-derived costimulatory ligand that is sufficient to induce a program of downstream events leading to T cell activation, breakdown of peripheral tolerance and induction of autoimmunity (summarized in Fig. 3). Although LIGHT can potentially bind three receptors,[9,16] HVEM is probably the receptor responsible for T-T cell interaction as LTβR is not found on T cells[4,12] and DcR3/TR6 is a decoy receptor that lacks a transmembrane domain.[16] Certainly, it is possible that LIGHT may have an unidentified receptor expressed on T cells. Moreover, a recent study suggests that LIGHT, although a ligand, can receive costimulatory signal when expressed on the T cell surface.[58] Due to the upregulation of LIGHT upon T cell activation, the simultaneous presence of both the ligand and receptor could provide a stimulatory mechanism for the clonal expansion of peripheral T cells in an autocrine or paracrine fashion as LIGHT can be secreted. Therefore, LIGHT, as a costimulatory molecule, basically causes differential responses for immature versus mature T cells. Upregulation of LIGHT promotes the deletion of potentially autoreactive T cells in thymic selection but activates mature T cells in periphery leading to autoimmune diseases.

Table 1. Complementation of $LT\alpha^{-/-}$ mice with LIGHT transgene in a LTβR-dependent manner[a]

	WT	$LT\alpha^{-/-}$	Tg $LT\alpha^{-/-}$	Tg $LT\beta R^{-/-}$
SLC	++	-	++	-
T/B cell segregation	++	-	++	-
$CD11c^+$ DC	++	±	++	±
$CR\text{-}Fc^+$ DC	++	-	+	-
FDC network	++	-	+	-
GC formation	++	-	+	-
Marginal Zone	++	-	-	-
KLH IgG response	++	-	++	-
SRBC IgG response	++	-	-	-

a) ++: strong positive; +: positive; ±: weak positive; -: undetectable.

LIGHT is a unique pro-inflammatory cytokine that not only effectively regulates T lymphocytes activation and effector function but also exerts its action on LTβR of stromal cells to mediate the formation of lymphoid structure in the absence of LT.[49] Highly organized lymphoid structures provide the intricate microenvironment essential for the mediation of the effective immune responses. Compared with $LT\beta^{-/-}$ mice, $LT\beta R^{-/-}$ mice present with more severely disorganized splenic structures, suggesting the potential involvement of another ligand.[5,59] We show that the complementation of $LT\alpha^{-/-}$ mice with a LIGHT transgene (LIGHT Tg/$LT\alpha^{-/-}$) leads to the restoration of secondary lymphoid-tissue chemokine (SLC) and T/B cell zone segregation (summarized in Table 1). LIGHT Tg/$LT\alpha^{-/-}$ mice also preserve dendritic cells (DC), follicular dendritic cell networks (FDC), and germinal centers (GC), though not the marginal zone (MZ). Consequently, IgG responses to soluble (KLH), but not particulate (SRBC), antigens are restored, confirming the differential role of primary follicle and marginal zone in the responses to soluble and particulate antigens. The failure of the LIGHT transgene to rescue the defective splenic structures in $LT\beta R^{-/-}$ mice demonstrates that LIGHT can interact with LTβR in vivo. These findings uncover the potential interaction between LIGHT and one of its receptors, LTβR, in supporting even in the absence of LT the development and maintenance of lymphoid microenvironment.[49] In addition, LTβR is essential for the development of secondary lymphoid organs such as lymph nodes (LN) and PP.[5] LIGHT has been identified as a ligand for LTβR in vitro[9] and cooperated with LTβ, another ligand for LTβR, in mesenteric lymph node (MLN) organogenesis.[33]

Both TNF and LT play critical roles in lymphoid neo-organogenesis and chronic inflammation as demonstrated in transgenic system.[2,8,60,61] Together with our study that LIGHT transgene can restore the formation of lymphoid tissue independent from LT, we propose that the overstimulation of either TNFR by TNF or LTβR by LIGHT under overexpression status may compensate the lack of LTβR signaling by membrane LTαβ. During local immune response or inflammation, high expression of LIGHT could have its potential to provide strong signal to form lymphoid-like structures. Therefore, LIGHT and LT can both play important roles in the formation of lymphoid tissues, though membrane LT appears to be dominant. Thus, the combined treatment of soluble receptors and antibodies may be required to block multiple ligands necessary for the formation of lymphoid structures or chemokine gradient during chronic inflammation.

T cells that mediate inflammation in a number of the experimental models have to migrate from sites of sensitization to sites of effector function to initiate and/or perpetuate the inflammatory response. Such migration is directed and depends on interaction between tissue-specific integrins and addressins and chemokine gradients, which in the case of traffic to mucosal tissues involves interactions between circulating cells bearing the α4β7 integrin and the MAdCAM-1 integrin on surface of endothelial cells.[62,63] Consistent with this possibility, MAdCAM-1 function has been shown to be critical to the development of colitis in the $CD45RB^{high}$ T cells transfer model.[64] Thus, molecules that may be relevant to recruitment and/or retention of cells within mucosal tissues can contribute to the mucosal inflammation. The ability of LIGHT inducing the production of chemokine and development of lymphoid structure may serve as an alternative mechanism by which upregulation of LIGHT can attract more T cells migrating into the local inflammation site and promote the transformation of the gut mucosa from a normal tertiary immune structure into a pathological lymphoid site. Similarly, local expression of LIGHT in tumor nodules leads to tumor rejection, probably due to both enhanced CTL response[11] and increased migration of T cells into tumor mediated by upregulation of chemokines and adhesion molecules inside tumor.

Summary

The unique features of LIGHT are still a matter of intense investigation. As we have known that LIGHT can function as a costimulatory molecule for T cells and promote the activation and expansion of T cells presumably by interacting with HVEM expressed on T cells.[11,13,30] LIGHT, cooperating with membrane LTαβ, plays an essential role in mesenteric lymph node organogenesis.[33] Moreover, LIGHT transgene can support the development and maintenance of lymphoid microenvironment independent from LT.[49] We propose that LIGHT plays a unique role in two key checkpoints for autoimmunity: activating autoreactive lymphocytes and also promoting the tissue infiltration of autoreactive T cells. Recent studies provide compelling evidence that LIGHT plays a critical role in T cell-mediated diseases including Type I diabetes, GVHD, IBD and tumor rejection. Thus, LIGHT may be an attractive candidate for therapeutic target and a better understanding of the mechanism(s) of its involvement in pathogenesis will allow us to develop effective treatment in the future.

References

1. Locksley RM, Killeen N, Lenardo MJ. The TNF and TNF receptor superfamilies: Integrating mammalian biology. Cell 2001; 104(4):487-501.
2. Fu YX, Chaplin DD. Development and maturation of secondary lymphoid tissues. Annu Rev Immunol 1999; 17:399-433.
3. Ettinger R. The role of tumor necrosis factor and lymphotoxin in lymphoid organ development. Curr Top Microbiol Immunol 2000; 251:203-210.
4. Ware CF, VanArsdale TL, Crowe PD et al. The ligands and receptors of the lymphotoxin system. Curr Top Microbiol Immunol 1995; 198:175-218.
5. Futterer A, Mink K, Luz A et al. The lymphotoxin β receptor controls organogenesis and affinity maturation in peripheral lymphoid tissues. Immunity 1998; 9:59-70.
6. Rennert PD, James D, Mackay F et al. Lymph node genesis is induced by signaling through the lymphotoxin β receptor. Immunity 1998; 9:71-79.
7. Wang Y, Wang J, Sun Y et al. Complementary effects of TNF and lymphotoxin on the formation of germinal center and follicular dendritic cells. J Immunol 2001; 166(1):330-337.
8. Ruddle NH. Lymphoid neo-organogenesis: Lymphotoxin's role in inflammation and development [In Process Citation]. Immunol Res 1999; 19(2-3):119-125.
9. Mauri DN, Ebner R, Montgomery RI et al. LIGHT, a new member of the TNF superfamily, and lymphotoxin alpha are ligands for herpesvirus entry mediator. Immunity 1998; 8(1):21-30.
10. Tamada K, Shimozaki K, Chapoval AI et al. LIGHT, a TNF-like molecule, costimulates T cell proliferation and is required for dendritic cell-mediated allogeneic T cell response. J Immunol 2000; 164(8):4105-4110.
11. Tamada K, Shimozaki K, Chapoval AI et al. Modulation of T-cell-mediated immunity in tumor and graft-versus-host disease models through the LIGHT costimulatory pathway. Nat Med 2000; 6(3):283-289.
12. Browning JL, Sizing ID, Lawton P et al. Characterization of lymphotoxin-alpha-beta complexes on the surface of mouse lymphocytes. J Immunol 1997; 159:3288-3298.
13. Harrop JA, McDonnell PC, Brigham-Burke M et al. Herpesvirus entry mediator ligand (HVEM-L), a novel ligand for HVEM/TR2, stimulates proliferation of T cells and inhibits HT29 cell growth. J Biol Chem 1998; 273(42):27548-27556.
14. Harrop JA, Reddy M, Dede K et al. Antibodies to TR2 (herpesvirus entry mediator), a new member of the TNF receptor superfamily, block T cell proliferation, expression of activation markers, and production of cytokines. J Immunol 1998; 161(4):1786-1794.
15. Kwon BS, Tan KB, Ni J et al. A newly identified member of the tumor necrosis factor receptor superfamily with a wide tissue distribution and involvement in lymphocyte activation. J Biol Chem 1997; 272(22):14272-14276.
16. Yu KY, Kwon B, Ni J et al. A newly identified member of tumor necrosis factor receptor superfamily (TR6) suppresses LIGHT-mediated apoptosis. J Biol Chem 1999; 274(20):13733-13736.
17. Granger SW, Butrovich KD, Houshmand P et al. Genomic characterization of LIGHT reveals linkage to an immune response locus on chromosome 19p13.3 and distinct isoforms generated by alternate splicing or proteolysis. J Immunol 2001; 167(9):5122-5128.
18. Flajnik MF, Kasahara M. Comparative genomics of the MHC: Glimpses into the evolution of the adaptive immune system. Immunity 2001; 15(3):351-362.
19. Rogers PR, Song J, Gramaglia I et al. OX40 promotes Bcl-xL and Bcl-2 expression and is essential for long-term survival of CD4 T cells. Immunity 2001; 15(3):445-455.

20. Kwon B, Lee HW, Kwon BS. New insights into the role of 4-1BB in immune responses: Beyond CD8+ T cells. Trends Immunol 2002; 23(8):378-380.
21. Lenschow DJ, Walunas TL, Bluestone JA. CD28/B7 system of T cell costimulation. Annu Rev Immunol 1996; 14:233-258.
22. Linsley PS, Brady W, Grosmaire L et al. Binding of the B cell activation antigen B7 to CD28 costimulates T cell proliferation and interleukin 2 mRNA accumulation. J Exp Med 1991; 173(3):721-730.
23. Razi-Wolf Z, Freeman GJ, Galvin F et al. Expression and function of the murine B7 antigen, the major costimulatory molecule expressed by peritoneal exudate cells. Proc Natl Acad Sci USA 1992; 89(9):4210-4214.
24. Reiser H, Freeman GJ, Razi-Wolf Z et al. Murine B7 antigen provides an efficient costimulatory signal for activation of murine T lymphocytes via the T-cell receptor/ CD3 complex. Proc Natl Acad Sci USA 1992; 89(1):271-275.
25. Sharon N. Lectin receptors as lymphocyte surface markers. Adv Immunol 1983; 34:213-298.
26. Ahmann GB, Sachs DH, Hodes RJ. Requirement for an Ia-bearing accessory cell in Con A-induced T cell proliferation. J Immunol 1978; 121(5):1981-1989.
27. Shahinian A, Pfeffer K, Lee KP et al. Differential T cell costimulatory requirements in CD28-deficient mice. Science 1993; 261(5121):609-612.
28. Perrin PJ, Davis TA, Smoot DS et al. Mitogenic stimulation of T cells reveals differing contributions for B7- 1 (CD80) and B7-2 (CD86) costimulation. Immunology 1997; 90(4):534-542.
29. Green JM, Noel PJ, Sperling AI et al. Absence of B7-dependent responses in CD28-deficient mice. Immunity 1994; 1(6):501-508.
30. Wang J, Lo JC, Foster A et al. The regulation of T cell homeostasis and autoimmunity by T cell-derived LIGHT. J Clin Invest 2001; 108(12):1771-1780.
31. Coyle AJ, Gutierrez-Ramos JC. The expanding B7 superfamily: Increasing complexity in costimulatory signals regulating T cell function. Nat Immunol 2001; 2(3):203-209.
32. Salomon B, Lenschow DJ, Rhee L et al. B7/CD28 costimulation is essential for the homeostasis of the CD4+CD25+ immunoregulatory T cells that control autoimmune diabetes. Immunity 2000; 12(4):431-440.
33. Scheu S, Alferink J, Potzel T et al. Targeted disruption of LIGHT causes defects in costimulatory T cell activation and reveals cooperation with lymphotoxin beta in mesenteric lymph node genesis. J Exp Med 2002; 195(12):1613-1624.
34. Tamada K, Ni J, Zhu G et al. Cutting edge: Selective impairment of CD8+ T cell function in mice lacking the TNF superfamily member LIGHT. J Immunol 2002; 168(10):4832-4835.
35. Shaikh RB, Santee S, Granger SW et al. Constitutive expression of LIGHT on T cells leads to lymphocyte activation, inflammation, and tissue destruction. J Immunol 2001; 167(11):6330-6337.
36. Allen JM, Forbush KA, Perlmutter RM. Functional dissection of the lck proximal promoter. Mol Cell Biol 1992; 12(6):2758-2768.
37. Greaves DR, Wilson FD, Lang G et al. Human CD2 3'-flanking sequences confer high-level, T cell-specific, position-independent gene expression in transgenic mice. Cell 1989; 56(6):979-986.
38. Gross JA, Johnston J, Mudri S et al. TACI and BCMA are receptors for a TNF homologue implicated in B-cell autoimmune disease. Nature 2000; 404(6781):995-999.
39. Mackay F, Woodcock SA, Lawton P et al. Mice transgenic for BAFF develop lymphocytic disorders along with autoimmune manifestations. J Exp Med 1999; 190(11):1697-1710.

40. Ware CF. APRIL and BAFF connect autoimmunity and cancer. J Exp Med 2000; 192(11):F35-38.
41. Kinne RW, Brauer R, Stuhlmuller B et al. Macrophages in rheumatoid arthritis. Arthritis Res 2000; 2(3):189-202.
42. Mahida YR. The key role of macrophages in the immunopathogenesis of inflammatory bowel disease. Inflamm Bowel Dis 2000; 6(1):21-33.
43. Datta SK, Patel H, Berry D. Induction of a cationic shift in IgG anti-DNA autoantibodies. Role of T helper cells with classical and novel phenotypes in three murine models of lupus nephritis. J Exp Med 1987; 165(5):1252-1268.
44. Granger SW, Ware CF. Turning on LIGHT. J Clin Invest 2001; 108(12):1741-1742.
45. Delovitch TL, Singh B. The nonobese diabetic mouse as a model of autoimmune diabetes: Immune dysregulation gets the NOD. Immunity 1997; 7(6):727-738. [published erratum appears in Immunity 1998; 8(4):531].
46. Tisch R, McDevitt H. Insulin-dependent diabetes mellitus. Cell 1996; 85(3):291-297.
47. Wu Q, Salomon B, Chen M et al. Reversal of spontaneous autoimmune insulitis in nonobese diabetic mice by soluble lymphotoxin receptor. J Exp Med J 2001; 193(11):1327-1332.
48. Ettinger R, Munson SH, Chao CC et al. A critical role for lymphotoxin-beta receptor in the development of diabetes in nonobese diabetic mice. J Exp Med 2001; 193(11):1333-1340.
49. Wang J, Foster A, Chin R et al. The complementation of lymphotoxin deficiency with LIGHT, a newly discovered TNF family member, for the restoration of secondary lymphoid structure and function. Eur J Immunol 2002; 32(7):1969-1979.
50. Tamada K, Tamura H, Flies D et al. Blockade of LIGHT/LTbeta and CD40 signaling induces allospecific T cell anergy, preventing graft-versus-host disease. J Clin Invest 2002; 109(4):549-557.
51. Ye Q, Fraser CC, Gao W et al. Modulation of LIGHT-HVEM costimulation prolongs cardiac allograft survival. J Exp Med 2002; 195(6):795-800.
52. Zhang J, Salcedo TW, Wan X et al. Modulation of T-cell responses to alloantigens by TR6/DcR3. J Clin Invest 2001; 107(11):1459-1468.
53. Blumberg RS, Saubermann LJ, Strober W. Animal models of mucosal inflammation and their relation to human inflammatory bowel disease. Curr Opin Immunol 1999; 11(6):648-656.
54. Kontoyiannis D, Pasparakis M, Pizarro TT et al. Impaired on/off regulation of TNF biosynthesis in mice lacking TNF AU-rich elements: Implications for joint and gut-associated immunopathologies. Immunity 1999; 10(3):387-398.
55. Stuber E, Strober W, Neurath M. Blocking the CD40L-CD40 interaction in vivo specifically prevents the priming of T helper 1 cells through the inhibition of interleukin 12 secretion. J Exp Med 1996; 183(2):693-698.
56. Wang J, Chun T, Lo JC et al. The critical role of LIGHT, a TNF family member, in T cell development. J Immunol 2001; 167(9):5099-5105.
57. Sebzda E, Mariathasan S, Ohteki T et al. Selection of the T cell repertoire. Annu Rev Immunol 1999; 17:829-874.
58. Shi G, Luo H, Wan X et al. Mouse T cells receive costimulatory signals from LIGHT, a TNF family member. Blood 2002; 100(9):3279-3286.
59. Koni PA, Sacca R, Lawton P et al. Distinct roles in lymphoid organogenesis for lymphotoxin α and β revealed in lymphotoxin β-deficient mice. Immunity 1997; 6:491-500.
60. Picarella DE, Kratz A, Li C-b et al. Transgenic tumor necrosis factor (TNF)-α production in pancreatic islets leads to insulitis, not diabetes. J Immunol 1993; 150:4136-4150.

61. Sacca R, Cuff CA, Lesslauer W et al. Differential activities of secreted lymphotoxin-alpha(3) and membrane lymphotoxin-alpha(1)beta(2) In lymphotoxin-Induced inflammation - critical role of TNF receptor 1 signaling. J Immunol 1998; 160:485-491.
62. Brandtzaeg P, Baekkevold ES, Morton HC. From B to A the mucosal way. Nat Immunol 2001; 2(12):1093-1094.
63. Butcher EC, Picker LJ. Lymphocyte homing and homeostasis. Science 1996; 272(5258):60-66.
64. Picarella D, Hurlbut P, Rottman J et al. Monoclonal antibodies specific for beta 7 integrin and mucosal addressin cell adhesion molecule-1 (MAdCAM-1) reduce inflammation in the colon of scid mice reconstituted with CD45RBhigh CD4+ T cells. J Immunol 1997; 158(5):2099-2106.

CHAPTER 6

CD137 Pathway in Innate and Adaptive Immunity

Ryan A. Wilcox and Lieping Chen*

Abstract

CD137 is a member of the TNF receptor superfamily which may be induced on a variety of cells, including activated T lymphocytes, natural killer cells and dendritic cells. Studies performed both in vitro and in vivo have suggested that CD137 activation pathway is capable of regulating cellular and molecular components of both innate and adaptive immunity. As we will discuss, interaction between CD137 receptor and its ligand may be a critical link between innate and adaptive immunity. Not surprisingly then, this receptor/ligand pair represents an attractive target for the immunotherapy. While there is ample evidence indicating that CD137 signaling promotes the regression of established tumors in mouse models, recent studies also demonstrate the role of CD137 in the inhibition of systemic autoimmune diseases in some animal models. Manipulation of this pathway may represent a promising approach for treatment of cancer and other immunological diseases.

CD137 Receptor and Ligand: Genes Expression and Biochemistry

CD137 (4-1BB, ILA) is a 30 kDa type I transmembrane glycoprotein belonging to the TNFR superfamily. The gene for mouse CD137, originally cloned in 1989 from a T cell specific cDNA library using a modified differential screening procedure, is located on chromosome 4. The human homologue is 60% identical at the amino acid level with murine CD137 and is located on chromosome 1p36 within a cluster of TNFR superfamily members, including CD30 and OX40.[1,2] Interestingly, translocations in this region have been associated with various hematopoietic malignancies.[3]

*Corresponding Author: Lieping Chen—Johns Hopkins University School of Medicine, Department of Dermatology, 600 N. Wolfe Street, Jefferson 1-127A, Baltimore, Maryland 21205, USA. Email: lchen42@jhmi.edu

TNF Superfamily, edited by Sanjay Khare. ©2007 Landes Bioscience

Inducible expression of the transmembrane form of CD137 has been observed in a broad range of both myeloid and lymphoid cells. In addition to $CD4^+$ and $CD8^+$ T cells, NK cells, neutrophils/granulocytes, eosinophils and monocytes are also found to express CD137.[4-8] In most cell types examined thus far, CD137 is expressed in an activation-dependent manner. Constitutive expression of CD137, albeit in low level, has recently been found on dendritic cells.[9] Although less well characterized, CD137 has also been observed on a variety of nonhematopoietic cells, including endothelium and vascular smooth muscle cells, bronchial epithelium, chondrocytes, and an osteosarcoma cell line.[10-13] CD137 is not only found on the cell surface, but also be expressed in a soluble form.[14] The identification of an mRNA variant lacking exon VIII, which encodes the transmembrane domain, would suggest that soluble CD137 (sCD137) represent an alternative splice form.[14] However, as members of the TNFR superfamily could be proteolytically cleaved from the cell surface, the possibility that sCD137 represents a shed form of membrane CD137 could not be excluded. In fact, soluble forms of the TNF superfamily member LIGHT could result from both alternative splicing and the proteolytic cleavage of the full-length -transmembrane protein.[15] It should be noted that sCD137 has been detected in the supernatants of activated T cells and in patients suffering from chronic inflammatory conditions, including rheumatoid arthritis and multiple sclerosis.[16,17] Although the function of sCD137 is not fully understood, our preliminary work suggests that sCD137 may act as a decoy receptor to block the CD137-CD137L interaction (Wilcox et al, unpublished data).

Although the cytoplasmic domain of CD137 lacks kinase activity, a Cys-X-Cys-Pro motif capable of binding the *src* family kinase $p56^{lck}$ has been identified.[18] Although $p56^{lck}$ coimmunoprecipitates with CD137, the significance of this association is not yet clear, as CD137 itself is not thought to be phosphorylated and an alternative substrate for CD137-associated $p56^{lck}$ is not yet known. Like many TNFR superfamily members, the cytoplasmic domain of CD137 interacts with a number of TRAFs. TRAFs are a family of six adaptor proteins that link members of the TNFR superfamily to downstream signaling pathways, leading to MAP kinase and NF-κB activation. While both human and murine CD137 may interact with TRAFs 1 and 2, human CD137 may also associate with TRAF3.[21,22] Upon CD137 engagement and the subsequent recruitment of TRAF2, activation of the MAP kinase kinase kinase, apoptosis signal-regulating kinase-1 (ASK-1), leads to the eventual activation of the MAP kinases JNK and p38 in murine T cells.[21,22] Activation of NF-κB following ligand binding is also dependent upon TRAF2 recruitment.[20] A leucine-rich repeat (LRR-1) protein was recently shown to interact with the cytoplasmic domain and inhibit NF-κB activation following CD137 stimulation;[23] however, the significance of this finding is not yet clear. Studies performed using TRAF2 deficient T cells have confirmed that TRAF2 is essential for IL-2 production following CD137 costimulation.[19] The ability of CD137 to activate both NF-κB and the MAP kinase JNK, may explain the ability of CD137 to stimulate IL-2 production and T cell proliferation in a CD28-independent fashion. Furthermore, the presence of both NF-κB and c-Jun binding sites in the CD137 promoter region suggests that CD137 signaling may upregulate its expression via a positive feedback loop.

The physiologic ligand for CD137 (CD137L) is a 50 kDa type II transmembrane glycoprotein expressed by professional antigen-presenting cells (APC), including B cells, macrophages and dendritic cells.[24-26] However, CD137L expression may not be restricted to APC. In fact, CD137L expression has been observed on activated T cells, cardiac myocytes and carcinoma cells of epithelial origin.[27-29] In addition, high levels of soluble CD137L (sCD137L) have been detected in the sera of patients with various hematological malignancies.[28] The release of sCD137L from the surface of various lymphocytes and monocytic cells was inhibited by a matrix metalloproteinase inhibitor, suggesting that sCD137L is generated following its cleavage from the cell surface. Furthermore, sCD137L was capable of binding its receptor and costimulating cytokine production in peripheral T cells. Several members of the TNF/TNFR superfamilies are capable of bidirectional signaling through both their receptors and respective ligands. Reverse signaling through CD137L has been shown to inhibit T cell proliferation and promote the induction of apoptosis in these cells.[30] Furthermore, CD137L stimulation promotes the release of IL-8 from carcinoma cells,[30] costimulates the proliferation of anti-μ-primed B cells,[31] and promotes survival and cytokine secretion in human monocytes.[32-35] The nature of signaling pathway(s) utilized by CD137L remains to be characterized. Extracellular matrix components, including fibronectin and collagen VI, have been shown to bind murine CD137.[36,37] However, the significance of this finding is unclear, as the ability of these extracellular matrix components to stimulate CD137 signaling has not been demonstrated. Furthermore, CD137 binding to the extracellular matrix may not be conserved across species, as human CD137 has not been shown to bind components of the extracellular matrix.

Collectively, the broad tissue distribution of inducible CD137 and its ligand, the presence of both soluble and transmembrane forms, and the ability of this receptor/ligand pair to stimulate a variety of different cell types, suggests that interactions between CD137 and its ligand may play an important role in regulating both the innate and adaptive immune responses.

CD137 and Innate Immunity

Natural killer (NK) cells, so named because of their ability to lyse selected tumor cells in vitro, are an important component of the innate immune response to virally infected and transformed cells. NK cells may not only kill sensitive target cells in a perforin-dependent manner, but they also provide an important source of IFN-γ and TNF. These NK cell-associated functions are regulated by various inhibitory and stimulatory receptors. Therefore, CD137's expression on activated NK cells[6] suggests that CD137 may regulate NK cell function. Although CD137 may not be capable of stimulating NK cell cytotoxicity, both CD137L-tranfectants and CD137 monoclonal antibodies (mAb) stimulated NK cell proliferation and IFN-γ secretion in vitro.[38] Furthermore, tumor eradication following CD137 mAb administration is NK cell-dependent, even though the tumors used were resistant to NK-mediated cytotoxicity. Finally, depletion of NK cells eliminated the ability of CD137 mAb to induce CTL generation.[6,39] Our results thus support the role of CD137 as a critical link between innate and adaptive immune responses.

Dendritic cells (DC) are an important component of innate immunity, as they process and present antigens and induce potent primary T cell responses. This is largely due to expression of abundant costimulatory molecules and an array of cytokines that are important in T cell activation. Recent work demonstrating CD137's constitutive expression on splenic DCs suggests that this receptor may be capable of regulating DC function.[9] In fact, coincubation of CD137L transfected tumor cells with DC induces secretion of IL-6 and IL-12 secretion.[9] Furthermore, splenic DCs isolated from mice given a CD137 mAb were better able to stimulate proliferation of antigen-specific T cells when compared to DCs isolated from mice which had received a control antibody.[9,40] Collectively, this data has implicated CD137 as an important receptor capable of stimulating DC maturation. Furthermore, the systemic administration of CD137 mAb in RAG-1-deficient mice was shown to enhance the ability of CD137-expressing splenic DCs isolated from these mice to stimulate T cell proliferation, providing direct evidence that activation of DC through CD137 may modulate DC function in vivo.

Monocytes and macrophages are important phagocytic cells which, upon migrating to an inflammatory site, secrete a wide variety of different cytokines, including IL-1, IL-6, IL-8, IL-12 and TNF, which have both local and systemic effects and serve to not only amplify the innate immune response but to trigger adaptive immunity. Not surprisingly, the accumulation of macrophages at sites of chronic inflammation is not uncommon and is regulated by chemokines that regulate cell migration, and by cytokines that prolong the survival of these cells. The recent observation that CD137, upon binding its ligand on human monocytes, is capable of stimulating the proliferation of these cells suggests that CD137/CD137L interactions may play an important role in the expansion of macrophages at inflammatory sites.[33] Furthermore, CD137L stimulation of monocytes has also been shown to stimulate the release of proinflammatory cytokines, including IL-6, IL-8 and TNF, and upregulates expression of the adhesion molecule ICAM-1.[32,34,35] Interestingly, CD137-stimulated monocytes were found to promote B cell apoptosis in a cell-contact dependent fashion, suggesting that CD137/CD137L interactions may inhibit the humoral immune response.[5] This observation is consistent to the recent finding that B cells were progressively deleted in CD137L transgenic mice in which the expression of CD137L is under the control of a MHC class II promoter.[41]

Granulocytes, including neutrophils and eosinophils, are important effector cells in the innate immune response to bacterial, fungal and parasitic pathogens. The accumulation of granulocytes at sites of inflammation is thought to be regulated by different cytokines, like G-CSF and GM-CSF for neutrophils and IL-5 for eosinophils, which inhibit apoptosis of these cells.[7] Both neutrophils and eosinophils were found to express CD137.[4,8] Interestingly, CD137 expression on eosinophils was only observed in patients suffering from IgE-mediated allergic responses, but not in normal subjects or those patients suffering from nonIgE-mediated eosinophilic disorders.[4] In both neutrophils and eosinophils, CD137 stimulation promoted apoptosis in these cells in the presence of GM-CSF and/or IL-5.[7] Therefore, CD137 stimulation may play an important role in regulating granulocyte survival during the initiation and resolution of an inflammatory response. Further studies, performed in CD137- or CD137L-deficient mice, will be required to demonstrate CD137's role in regulating monocyte or granulocyte function during an innate immune response in vivo.

CD137 and Adaptive Immunity

Studies performed in vitro utilizing either CD137L-transfected cells or agonistic anti-CD137 mAb have shown that CD137 is capable of stimulating proliferation and IL-2 secretion in both $CD4^+$ and $CD8^+$ T cells.[38,39] However, CD137's ability to costimulate T cell proliferation may not be entirely attributed to its ability to stimulate IL-2 secretion, as CD137 mAb enhanced the proliferation of $CD8^+$ T cells isolated from IL-2-deficient mice in an allogeneic mixed lymphocyte reaction.[44] In addition to its ability to stimulate T cell proliferation and cytokine secretion, CD137 may also prevent activation induced cell death, as CD137 mAb administration inhibited the deletion of superantigen-activated T cells.[43,45] In a similar fashion, the ability of LPS administration to prevent the deletion of superantigen-activated T cells was partially abrogated upon the administration of a CD137Ig fusion protein.[43] Therefore, endogenous CD137L may promote the survival of activated T cells. It should be noted that the ability of CD137 costimulation to promote the survival of activated T cells was most pronounced in $CD8^+$ cells.

In addition to promoting cell expansion and survival, CD137 also stimulates the production of effector cytokines, most noticeably IFN-γ.[46,47] In order to determine the importance of endogenous CD137L in the generation of IFN-γ-producing cells, ovalbumin (OVA)-specific $CD8^+$ TCR transgenic T cells isolated from OT-1 mice were adoptively transferred into wild type recipients. Recipient mice were subsequently immunized with OVA and given either a control fusion protein or a CD137 fusion protein (CD137Ig). Upon restimulation in vitro, OT-1 cells isolated from the CD137Ig-treated mice secreted 50% less IFN-γ than those T cells isolated from the control mice, even though the development of cytotoxicity was unimpaired.[47] Similarly, administration of an antagonistic CD137L mAb ameliorated acute graft versus host diseases (GVHD), but promoted autoantibody production in chronic GVHD.[48] This may be attributed, at least in part, to the ability of CD137L mAb administration to inhibit both the expansion and IFN-γ production of alloreactive $CD8^+$ T cells in recipient mice, thus promoting the development of a type 2 response in chronic GVHD.

Although CD137 costimulation clearly promotes the expansion and survival of CD8$^+$ T cells, CD137's role in the CD4$^+$ T cell response is less clear. Cannons et al demonstrated that CD137L was capable of promoting the expansion and survival of both CD4$^+$ and CD8$^+$ T cells in vitro.[42] Furthermore, Blazar et al demonstrated, using both CD137-deficient mice and CD137 mAb, that CD137 costimulation was capable of stimulating GVHD mediated by either CD4$^+$ or CD8$^+$ T cells.[49] In contrast, experimental autoimmune encephalomyelitis (EAE), a Th1-mediated autoimmune response, was ameliorated upon CD137 mAb administration.[50] Although CD137 mAb administration promoted autoimmune CD4$^+$ T-cell expansion initially, these T cells undergo programmed cell death, suggesting that CD137 may actually stimulate activation induced cell death in this setting. In addition, the same agonistic CD137 mAb also prevented and inhibited progressive lymphoprolifeative diseases and lupus-like symptoms in Fas-deficient MRL/lpr mice accompanied with profound deletion of autoreactive B cells and double negative T cells.[51] These results are consistent with a previous report demonstrating the ability of CD137 mAb to promote activation induced cell death (AICD) in several in vitro-stimulated CD4$^+$ T cell clones.[25] These findings, however, also indicate that Fas is not involved in AICD in these systems, although it was reported that FasL is upregulated on CD4$^+$ T cells following CD137 costimulation.[52]

Endogenous CD137 activation does not appear to be important in the generation of normal humoral immune responses. For example, CD137L-deficient mice have a normal number and distribution of lymphocytes and were capable of generating a normal level of IgG antibodies following LCMV, VSV or influenza virus infection.[53-55] However, administration of agonistic CD137 mAb inhibited the development of a humoral immune response to T cell dependent antigens.[56] Adoptive transfer experiments suggest that the ability of CD137 stimulation to inhibit the antibody response is mediated by CD137-stimulated CD4$^+$ T cells. A role for CD137 signaling in B cells was not observed.

Studies performed in both CD137- and CD137L-deficient mice have confirmed the importance of CD137 costimulation in the generation of a fully competent T cell response. While the CD4$^+$ T cell response was undiminished in knockout mice following LCMV.[54] or influenza virus infection,[57] the expansion of virus-specific CD8$^+$ T cells was significantly reduced. Similarly, the VSV-specific CTL response was reduced in CD137-deficient mice.[58] It should be noted, however, that the T cells isolated from the CD137-deficient mice exhibited an enhanced proliferative response following anti-CD3 stimulation.[58] While the significance of this observation is not fully understood, the absence of CD137, whether in its cell-bound or soluble forms, may eliminate a pro-apoptotic signal through CD137L, thus promoting T cell expansion in vitro.[30] The generation of LCMV-specific CD8$^+$ T cells was greatly reduced in CD137L-deficient mice following peptide immunization.[55] While the wild type mice were protected against a LCMV challenge following peptide immunization, the ability to clear the virus was reduced in the CD137L-deficient mice. More recent experiments have analyzed both the primary and secondary CTL response following influenza infection. While the primary CTL response was undiminished in the CD137L-deficient mice, the memory response was markedly reduced in these mice.[57] Collectively, these studies suggest that CD137L, while unnecessary for the generation of a $CD4^+$ T cell response, is required for the generation of a fully competent $CD8^+$ T cell response. Given the importance of CD137 costimulation in the generation of a fully competent T cell response, CD137 and its ligand represent attractive targets in the immunotherapy setting. In fact, manipulation of this costimulatory pathway has been utilized for treatment of acute GVHD,[48] EAE,[50] lymphoproliferative diseases,[51] myocarditis,[27] viral infections[59] and cancers in animal models.[60-65] In the following section, we will focus our discussion to CD137 manipulation in the potentiation of cancer immunity.

CD137 and Tumor Immunotherapy

Melero and colleagues were the first to demonstrate that CD137 stimulation, provided by a CD137 mAb, greatly enhanced the tumor-specific CTL response and eradicated established tumors in mouse models.[61] This particular mAb, 1D8, costimulated T cell growth and cytokine release when provided in immobilized, but not soluble form, in the presence of immobilized anti-CD3 mAb as TCR signal, suggesting that this mAb delivers agonistic signaling for T cells. Tumor eradication following CD137 mAb administration in the P815 mastocytoma model used in this study was dependent upon both $CD4^+$ and $CD8^+$ T cells and NK cells. The antitumor effect was less dependent on $CD4^+$ T cells as depletion of $CD4^+$ T cells by specific mAb only partially eliminated the antitumor effect of 1D8, and depletion of $CD8^+$ T cells or NK cells completely abrogated the effect.[6,39] Interestingly, a poorly immunogenic tumor such as AG104A sarcoma was also sensitive to the CD137 mAb treatment.[61] Further studies demonstrate, however, several poorly immunogenic tumors were resistant to treatment.[39,65] For example, a T cell response was undetectable in mice during progressive growth of an epithelial-derived C3 tumor expressing the E7 oncogene of human papillomavirus (HPV) type 16 [39] Immunization with a peptide encoding a $H\text{-}2D^b$ restricted epitopes of the E7 protein, however, induced normal T cell responses in tumor-bearing mice, suggesting that the tumor may be ignored by the host immune system.[39] Not surprisingly, C3 tumor regression was not observed in tumor-bearing mice following CD137 mAb treatment, although antibody administration following immunization with the E7 peptide generated a tumor-specific CTL response capable of eradicating established tumors in the majority of mice. Kim et al demonstrated that while CD137 mAb administration prolonged survival in mice bearing established intracranial (i.c.) tumors, the same tumors were resistant to treatment when grown subcutaneously (s.c.). However, the s.c. tumors regressed following the concomitant treatment of an i.c. tumor.[65] The demonstration of OX-40^+ cells within the i.c. tumors suggests that these tumors were capable of eliciting an immune response. These results thus suggest that the tumors grown s.c. are ignored and thus resistant to treatment with CD137 mAb alone, although the same tumors i.c. could prime T cell responses. While many costimulatory receptors are constitutively expressed on naïve T cells, CD137 is inducibly expressed following T cell activation. This may explain the inability of CD137 mAb administration to eradicate poorly immunogenic tumors that fail to prime a tumor-specific T cell response. A direct implication from these studies is that CD137 mAb administration may represent an effective form of immunotherapy for those tumors that elicit a T cell response, albeit it may be weak one. However, those tumors that are immunologically ignored may only regress when CD137 mAb is used as an adjunct with antigen-based forms of immunotherapy. Other methods that could stimulate priming of T cells may increase the effect of CD137 mAb. For example, CD137 mAb administration was shown to be effective in a poorly immunogenic colon carcinoma model following the intratumoral delivery of a recombinant adenovirus expressing IL-12.[62] The NK cell response generated following the adenoviral-mediated expression of IL-12 may not only promote the cross-presentation of tumor antigens required for the generation of a tumor-specific CTL response, but also play an immunoregulatory role required for CTL generation.[66] Similar results were also obtained in this model following the intratumoral delivery of recombinant adenoviruses expressing IL-12 and CD137L, further supporting the notion that CD137L may represent an attractive target for gene therapy.[63] Alternatively, gene transfer of a single-chain Fv fragment specific for CD137 was recently shown to stimulate a $CD4^+$ T cell response against a poorly immunogenic melanoma cell.[64] While CD137 is generally thought to stimulate a $CD8^+$ T cell response in the tumor-bearing host, in this model, tumor eradication was dependent upon both NK cells and $CD4^+$ T cells, although a role for $CD8^+$ T cells was not observed. Whether or not CD137 stimulates NK-cell cytokine secretion and/or cytotoxicity in this model remains to be shown. Because K1735 tumor is sensitive to NK-mediated lysis, it seems likely that NK cells are the major effector cells and activation of $CD4^+$ T cells promotes NK activity.[67]

Although there is clear evidently that CD137 may be manipulated for potentiation of tumor immunity, much less is known about the mechanism involved. As mentioned, $CD4^+$[61,64] and $CD8^+$ T cells,[39,41-63] as well as NK cells,[6,39,62,63] have been implicated. Furthermore, regulation of DC function by CD137 signaling suggests that these professional APC may be involved in either the cross-presentation of exogenous tumor antigens or the direct-presentation of endogenous antigens following CD137 stimulation.[9,36,64] Several mechanisms, none of which are mutually exclusive, may explain the ability of CD137 mAb, for example, to stimulate an anti-tumor immune response. First, CD137 signaling stimulates cytokine production (e.g., IL-12) on dendritic cells,[39,40] thus promoting their ability to stimulate a productive T cell response. Furthermore, CD137 costimulation could stimulate a $CD8^+$ T cell response against an MHC class I-deficient tumor and bypass the need for $CD4^+$ T-cell help.[68] Whether or not CD137-stimulated DCs are involved remains to be shown. CD137 may also stimulate proliferation and cytokine secretion in tumor-specific $CD4^+$ T cells. Furthermore, CD137 signaling may upregulate FasL expression on $CD4^+$ effector T cells, thus promoting cytotoxicity.[52] While a $CD4^+$ T cell response may be required for tumor eradication following CD137 stimulation in some tumor models, tumor eradication is CD4-independent in others. In fact, depletion of $CD4^+$ T cells appeared to further improve the response rate following CD137 mAb administration in one tumor model.[39] Whether or not this may be attributed to a loss of $CD4^+CD25^+$ suppressor cells remains to be shown.[69,70] CD137 may not only stimulate the expansion of tumor-specific CTL, but also promote their survival. Furthermore, CD137 may be important for the development of IFN-γ-producing effector cells.[47] Tumor eradication following CD137 mAb administration was IFN-γ dependent, and although IFN-γ was not required for the differentiation of cytotoxic T cells, the accumulation of tumor-specific CTL was impaired in the absence of IFN-γ.[71] Whether or not this may be attributed to the diminished production of IFN-γ inducible chemokines (e.g., Mig, IP-10, I-TAC) at the tumor site remains to be shown. In addition to direct triggering of T cells, CD137 signaling may also stimulate proliferation and cytokine secretion in activated NK cells, leading to the development of immune regulatory function. This may be supported by the observation that tumor-specific CTL activity following CD137 mAb administration was diminished in NK cell depleted mice.[6,39] Experiments performed in CD137-deficient mice should help to further clarify the contributions made by various cell subsets following CD137 stimulation.

Concluding Remarks

CD137 and its ligand are expressed on a wide variety of cell types in both transmembrane and soluble forms. Signaling through CD137 is capable of activating DC, NK and T cells and provides a critical link between innate and adaptive immunity. Understanding the mechanisms underlying CD137's role in immunity may lead to the development of new and improved forms of immunotherapy for a broad spectrum of disease states, including the prevention of autoimmunity or graft rejection and potentiation of a vigorous tumor-specific immune response.

References

1. Pollok KE, Kim YJ, Zhou Z et al. Inducible T cell antigen 4-1BB. Analysis of expression and function. J Immunol 1993; 150:771-81.
2. Kwon BS, Kozak CA, Kim KK et al. Genomic organization and chromosomal localization of the T-cell antigen 4-1BB. J Immunol 1994; 152:2256-62.
3. Schwarz H, Arden K, Lotz M. CD137, a member of the tumor necrosis factor receptor family, is located on chromosome 1p36, in a cluster of related genes, and colocalizes with several malignancies. Biochem Biophys Res Commun 1997; 235:699-70.
4. Heinisch IV, Bizer C, Volgger W et al. Functional CD137 receptors are expressed by eosinophils from patients with IgE-mediated allergic responses but not by eosinophils from patients with nonIgE-mediated eosinophilic disorders. J Allergy Clin Immunol 2001; 108:21-8.

5. Kienzle G, von Kempis J. CD137 (ILA/4-1BB), expressed by primary human monocytes, induces monocyte activation and apoptosis of B lymphocytes. Int Immunol 2000; 12:73-82.
6. Melero I, Johnston JV, Shufford WW et al. NK1.1 cells express 4-1BB (CDw137) costimulatory molecule and are required for tumor immunity elicited by anti-4-1BB monoclonal antibodies. Cell Immunol 1998; 190:167-72.
7. Simon HU. Evidence for a pro-apoptotic function of CD137 in granulocytes. Swiss Med Wkly 2001; 131:455-8.
8. Heinisch IV, Daigle I, Knopfli B.CD137 activation abrogates granulocyte-macrophage colony-stimulating factor-mediated anti-apoptosis in neutrophils. Eur J Immunol 2000; 30:3441-6.
9. Wilcox RA, Chapoval AI, Gorski KS et al. Cutting Edge: Expression of functional CD137 receptor by dendritic cells. J Immunol 2002; 168:4262-7.
10. Broll K, Richter G, Pauly S et al. CD137 expression in tumor vessel walls. High correlation with malignant tumors. Am J Clin Pathol 2001; 115:543-9.
11. Boussaud V, Soler P, Moreau J et al. Expression of three members of the TNF-R family of receptors (4-1BB, lymphotoxin-beta receptor, and Fas) in human lung. Eur Respir J 1998; 12:926-31.
12. von Kempis J, Schwarz H, Lotz M. Differentiation-dependent and stimulus-specific expression of ILA, the human 4-1BB-homologue, in cells of mesenchymal origin. Osteoarthritis Cartilage 1997; 5:394-406.
13. Lisignoli G, Toneguzzi S, Cattini L et al. Different expression pattern of cytokine receptors by human osteosarcoma cell lines. Int J Oncol 1998; 12:899-903.
14. Setareh M, Schwarz H, Lotz M. A mRNA variant encoding a soluble form of 4-1BB, a member of the murine NGF/TNF receptor family. Gene 1995; 164:311-5.
15. Granger SW, Butrovich KD, Houshmand P et al. Genomic characterization of LIGHT reveals linkage to an immune response locus on chromosome 19p13.3 and distinct isoforms generated by alternate splicing or proteolysis. J Immunol 2001; 167:5122-8.
16. Michel J, Langstein J, Hofstadter F et al. A soluble form of CD137 (ILA/4-1BB), a member of the TNF receptor family, is released by activated lymphocytes and is detectable in sera of patients with rheumatoid arthritis. Eur J Immunol 1998; 28:290-5.
17. Sharief MK. Heightened intrathecal release of soluble CD137 in patients with multiple sclerosis. Eur J Neurol 2002; 9:49-54.
18. Kim YJ, Pollok KE, Zhou Z et al. Novel T cell antigen 4-1BB associates with the protein tyrosine kinase p56lck1. J Immunol 1993; 151:1255-62.
19. Saoulli K, Lee SY, Cannons JL et al. CD28-independent, TRAF2-dependent costimulation of resting T cells by 4- 1BB ligand. J Exp Med 1998; 187:1849-62.
20. Jang IK, Lee ZH, Kim YJ et al. Human 4-1BB (CD137) signals are mediated by TRAF2 and activate nuclear factor-kappa B. Biochem Biophys Res Commun 1998; 242:613-20.
21. Cannons JL, Choi Y, Watts TH. Role of TNF receptor-associated factor 2 and p38 mitogen-activated protein kinase activation during 4-1BB-dependent immune response. J Immunol 2000; 165:6193-204.
22. Cannons JL, Hoeflich KP, Woodgett JR et al. Role of the stress kinase pathway in signaling via the T cell costimulatory receptor 4-1BB. J Immunol 1999; 163:2990-8.
23. Jang LK, Lee ZH, Kim HH et al. A novel leucine-rich repeat protein (LRR-1): Potential involvement in 4- 1BB-mediated signal transduction. Mol Cells 2001; 12:304-12.
24. Goodwin RG, Din WS, Davis-Smith T et al. Molecular cloning of a ligand for the inducible T cell gene 4-1BB: A member of an emerging family of cytokines with homology to tumor necrosis factor. Eur J Immunol 1993; 23:2631-41.

25. Alderson MR, Smith CA, Tough TW et al. Molecular and biological characterization of human 4-1BB and its ligand. Eur J Immunol 1994; 24:2219-27.
26. Zhou Z, Kim S, Hurtado J et al. Characterization of human homologue of 4-1BB and its ligand. Immunol Lett 1995; 45:67-73.
27. Seko Y, Takahashi N, Oshima H et al. Expression of tumour necrosis factor (TNF) ligand superfamily co stimulatory molecules CD30L, CD27L, OX40L, and 4-1BBL in murine hearts with acute myocarditis caused by Coxsackievirus B3. J Pathol 2001; 195:593-603.
28. Salih HR, Schmetzer HM, Burke C et al. Soluble CD137 (4-1BB) ligand is released following leukocyte activation and is found in sera of patients with hematological malignancies. J Immunol 2001; 167:4059-66.
29. Salih HR, Kosowski SG, Haluska VF et al. Constitutive expression of functional 4-1BB (CD137) ligand on carcinoma cells. J Immunol 2000; 165:2903-10.
30. Schwarz H, Blanco FJ, von Kempis J et al. ILA, a member of the human nerve growth factor/tumor necrosis factor receptor family, regulates T-lymphocyte proliferation and survival. Blood 1996; 87:2839-45.
31. Pollok KE, Kim YJ, Hurtado J et al. 4-1BB T-cell antigen binds to mature B cells and macrophages, and costimulates anti-mu-primed splenic B cells. Eur J Immunol 1994; 24:367-74.
32. Langstein J, Becke FM, Sollner L et al. Comparative analysis of CD137 and LPS effects on monocyte activation, survival, and proliferation. Biochem Biophys Res Commun 2000; 273:117-22.
33. Langstein J, Michel J, Schwarz H. CD137 induces proliferation and endomitosis in monocytes. Blood 1999; 94:3161-8.
34. Langstein J, Schwarz H. Identification of CD137 as a potent monocyte survival factor. J Leukoc Biol 1999; 65:829-33.
35. Langstein J, Michel J, Fritsche J et al. CD137 (ILA/4-1BB), a member of the TNF receptor family, induces monocyte activation via bidirectional signaling. J Immunol 1998; 160:2488-94.
36. Chalupny NJ, Peach R, Hollenbaugh D et al. T-cell activation molecule 4-1BB binds to extracellular matrix proteins. Proc Natl Acad Sci USA 1992; 89:10360-4.
37. Loo DT, Chalupny NJ, Bajorath J et al. Analysis of 4-1BBL and laminin binding to murine 4-1BB, a member of the tumor necrosis factor receptor superfamily, and comparison with human 4- 1BB. J Biol Chem 1997; 272:6448-56.
38. Wilcox RA, Tamada K, Strome SE et al. Signaling through NK cell-associated CD137 promotes both helper function for CD8+ cytolytic T lymphocytes and responsiveness to interleukin-2 but not cytolytic activity. J Immunol 2002; 169:4230-6.
39. Wilcox RA, Flies DB, Zhu G et al. Provision of antigen and CD137 signaling breaks immunological ignorance, promoting regression of poorly immunogenic tumors. J Clin Invest 2002; 109:651-9.
40. Futagawa T, Akiba H, Kodama T et al. Expression and function of 4-1BB and 4-1BB ligand on murine dendritic cells. Int Immunol 2002; 14:275-86.
41. Zhu G, Flies DB, Tamada K et al. Progressive depletion of peripheral B lymphocytes in 4-1BB (CD137) ligand/I-Eα)-transgenic mice. J Immunol 2001; 167:2671-6.
42. Cannons JL, Lau P, Ghumman B et al. 4-1BB ligand induces cell division, sustains survival, and enhances effector function of CD4 and CD8 T cells with similar efficacy. J Immunol 2001; 167:1313-24.
43. Takahashi C, Mittler RS, Vella AT. Differential clonal expansion of CD4 and CD8 T cells in response to 4- 1BB ligation: Contribution of 4-1BB during inflammatory responses. Immunol Lett 2001; 76:183-91.

44. Tan JT, Ha J, Cho HR et al. Analysis of expression and function of the costimulatory molecule 4-1BB in alloimmune responses. Transplantation 2000; 70:175-83.
45. Takahashi C, Mittler RS, Vella AT. Cutting edge: 4-1BB is a bona fide CD8 T cell survival signal. J Immunol 1999; 162:5037-40.
46. Shuford WW, Klussman K, Tritchler DD et al. 4-1BB costimulatory signals preferentially induce CD8+ T cell proliferation and lead to the amplification in vivo of cytotoxic T cell responses. J Exp Med 1997; 186:47-55.
47. Cooper D, Bansal-Pakala P, Croft M. 4-1BB (CD137) controls the clonal expansion and survival of CD8 T cells in vivo but does not contribute to the development of cytotoxicity. Eur J Immunol 2002; 32:521-9.
48. Nozawa K, Ohata J, Sakurai J et al. Preferential blockade of CD8(+) T cell responses by administration of anti-CD137 ligand monoclonal antibody results in differential effect on development of murine acute and chronic graft-versus-host diseases. J Immunol 2001; 167:4981-6.
49. Blazar BR, Kwon BS, Panoskaltsis-Mortari A et al. Ligation of 4-1BB (CDw137) regulates graft-versus-host disease, graft- versus-leukemia, and graft rejection in allogeneic bone marrow transplant recipients. J Immunol 2001; 166:3174-83.
50. Sun Y, Lin X, Chen HM et al. Administration of agonistic anti-4-1BB monoclonal antibody leads to the amelioration of experimental autoimmune encephalomyelitis. J Immunol 2002; 168:1457-65.
51. Sun Y, Chen HM, Subudhi SK et al. Costimulatory molecule-targeted antibody therapy of a spontaneous autoimmune disease. Nature Med 2002; 8:1405-1413.
52. Ebata T, Mogi S, Hata Y et al. Rapid induction of CD95 ligand and CD4+ T cell-mediated apoptosis by CD137 (4-1BB) costimulation. Eur J Immunol 2001; 31:1410-6.
53. DeBenedette MA, Wen T, Bachmann MF et al. Analysis of 4-1BB ligand (4-1BBL)-deficient mice and of mice lacking both 4-1BBL and CD28 reveals a role for 4-1BBL in skin allograft rejection and in the cytotoxic T cell response to influenza virus. J Immunol 1999; 163:4833-41.
54. Tan JT, Whitmire JK, Ahmed R et al. 4-1BB ligand, a member of the TNF family, is important for the generation of antiviral CD8 T cell responses. J Immunol 1999; 163:4859-68.
55. Tan JT, Whitmire JK, Murali-Krishna K et al. 4-1BB costimulation is required for protective anti-viral immunity after peptide vaccination. J Immunol 2000; 164:2320-5.
56. Mittler RS, Bailey TS, Klussman K et al. Anti-4-1BB monoclonal antibodies abrogate T cell-dependent humoral immune responses in vivo through the induction of helper T cell anergy. J Exp Med 1999; 190:1535-40.
57. Bertram EM, Lau P, Watts TH. Temporal segregation of 4-1BB versus CD28-mediated costimulation: 4-1BB ligand influences T cell numbers late in the primary response and regulates the size of the T cell memory response following influenza infection. J Immunol 2002; 168:3777-85.
58. Kwon BS, Hurtado JC, Lee ZH et al. Immune responses in 4-1BB (CD137)-deficient mice. J Immunol 2002; 168:5483-90.
59. Halstead ES, Mueller YM, Altman JD et al. In vivo stimulation of CD137 broadens primary antiviral CD8(+) T cell responses. Nat Immunol 2002; 3:536-41.
60. Melero I, Bach N, Hellstrom E et al. Amplification of tumor immunity by gene transfer of the costimulatory 4-1BB ligand: Synergy with the CD28 costimulatory pathway. Eur J Immunol 1998; 28:1116-21.
61. Melero I, Shuford WW, Newby SA et al. Monoclonal antibodies against the 4-1BB T-cell activation molecule eradicate established tumors. Nat Med 1997; 3:682-5.

62. Chen SH, Pham-Nguyen KB, Martinet O et al. Rejection of disseminated metastases of colon carcinoma by synergism of IL-12 gene therapy and 4-1BB costimulation. Mol Ther 2000; 2:39-46.
63. Martinet O, Ermekova V, Qiao JQ et al. Immunomodulatory gene therapy with interleukin 12 and 4-1BB ligand: Long- term remission of liver metastases in a mouse model. J Natl Cancer Inst 2000; 92:931-6.
64. Ye Z, Hellstrom I, Hayden-Ledbetter M et al. Gene therapy for cancer using single-chain Fv fragments specific for 4- 1BB. Nat Med 2002; 8:343-8.
65. Kim JA, Averbook BJ, Chambers K et al. Divergent effects of 4-1BB antibodies on antitumor immunity and on tumor-reactive T-cell generation. Cancer Res 2001; 61:2031-7.
66. Kurosawa S, Harada M, Matsuzaki G et al. Early-appearing tumour-infiltrating natural killer cells play a crucial role in the generation of anti-tumour T lymphocytes. Immunology 1995; 85:338-46.
67. Chen L. Antibody gene therapy: Old wine in a new bottle. Nature Med 2002; 8:333-4.
68. Diehl L, van Mierlo GJ, den Boer AT et al. In vivo triggering through 4-1BB enables Th-independent priming of CTL in the presence of an intact CD28 costimulatory pathway. J Immunol 2002; 168:3755-62.
69. Chatenoud L, Salomon B, Bluestone JA. Suppressor T cells—they're back and critical for regulation of autoimmunity! Immunol Rev 2001; 182:149-63.
70. Tanaka H, Tanaka J, Kjaergaard J et al. Depletion of CD4+ CD25+ regulatory cells augments the generation of specific immune T cells in tumor-draining lymph nodes. J Immunother 2002; 25:207-17.
71. Wilcox RA, Flies DB, Wang H et al. Impaired infiltration of tumor-specific cytolytic T cells in the absence of interferon-γ despite their normal maturation in lymphoid organs during anti-CD137 monoclonal antibody therapy. Cancer Res 2002; 62:4413-8.

Index

J

K

L

M

N

O

P

R

S

T

V

For Product Safety Concerns and Information please contact our EU
representative GPSR@taylorandfrancis.com Taylor & Francis Verlag GmbH,
Kaufingerstraße 24, 80331 München, Germany

Printed and bound by CPI Group (UK) Ltd, Croydon, CR0 4YY
14/09/2026
02216243-0018